T0221067

ELEMENTARY
NUMBER THEORY
WITH PROGRAMMING

ELEMENTARY NUMBER THEORY WITH PROGRAMMING

MARTY LEWINTER

JEANINE MEYER

Published by John Wiley & Sons, Inc., Hoboken, New Jersey
Published simultaneously in Canada

For general information on our other products and services or for technical support, please contact our Customer Care Department within the United States at (800) 762-2974, outside the United States at (317) 572-3993 or fax (317) 572-4002.

Wiley also publishes its books in a variety of electronic formats. Some content that appears in print may not be available in electronic formats. For more information about Wiley products, visit our web site at www.wiley.com.

Library of Congress Cataloging-in-Publication Data:

Lewinter, Marty, 1950–
  Elementary number theory with programming / Marty Lewinter, Jeanine Meyer.
    pages   cm
  Includes index.
    ISBN 978-1-119-06276-9 (cloth)
1. Number theory.   2. Number theory–Problems, exercises, etc.   3. Computer programming.
I. Meyer, Jeanine.   II. Title.   III. Title: Number theory with programming.
  QA241.L5815 2015
  512.7–dc23

                                                                              2015000699

Set in 11/13pt Times by SPi Global, Pondicherry, India

10   9   8   7   6   5   4   3   2   1

The first author dedicates this book to his son and fellow mathematician, Anthony Delgado

The second author dedicates this book to her mother, Esther Minkin, of blessed memory

CONTENTS

PREFACE

"Everything is number."

—Pythagoras, sixth century B.C.

A question is sometimes asked if mathematics is discovery or invention. Certainly many of the definitions were devised by mathematicians. However, the relationships of the terms and the characterizations are not arbitrary or purely descriptive but proven by logic.

The logic of mathematics and the logic of programming are similar, and improving skills in one will help the other. The beauty of a proof is similar to the beauty of a program.

Elementary number theory is a special branch of mathematics in that many of the proven theorems and many of the conjectures can be stated so that anyone with an elementary knowledge of algebra can understand them.

This textbook was developed to be used in a college mathematics and computer science program. However, it can be used at institutions with separate mathematics and computing majors. The material in this book is also suitable for a course at the high school level. The clever and esthetic argument drawn from this text will enhance a student's admiration for the power of high school algebra.

The first author fell in love with mathematics in the fifth grade. The teacher said that in order to divide by a fraction, we must multiply by its reciprocal. Thus, to divide 8 by 2/3, we multiply by 3/2, obtaining $8 \times 3/2 = 12$. When asked why this works, the teacher replied, "Just do it." That evening, the first author reasoned as follows. To divide

8 by 1/3, we clearly multiply by 3, since there are three "thirds" in each "one." Thus, if each guest eats one third of a pizza pie, then 8 pies can feed 8×3, or 24 guests. On the other hand, if each guest eats two thirds of a pie, that is, if each guest eats twice as much, then only half as many guests (12 guests) can be fed. Thus, to divide 8 by 2/3, that is, to determine how many "2/3 of a pie" there are in 8 pies, we wind up multiplying by 3 and dividing by 2. In other words, we multiply by 3/2.

Every integer has its secret! The cube 125 (i.e., $5 \times 5 \times 5$) is the sum of two squares in two different ways: $125 = 100 + 25 = 121 + 4$. The number 55 is the sum of the first 10 numbers: $1 + 2 + 3 + 4 + 5 + 6 + 7 + 8 + 9 + 10 = (1 + 10) + (2 + 9) + (3 + 8) + (4 + 7) + (5 + 6) = 5 \times 11 = 55$. The innocent-looking number, 16, is the only number that can be written as a^b and b^a where a and b are distinct positive integers: $16 = 2^4 = 4^2$. The list of wonders never ceases. Nor do the open questions that still challenge the entire world community of mathematicians. Is every even number except 2 the sum of two prime numbers? (A prime number is divisible only by itself and by 1.) We have $4 = 2 + 2$, $6 = 3 + 3$, $8 = 5 + 3$, $10 = 7 + 3$, $12 = 7 + 5$, but no one knows whether the general conjecture made over 200 years ago is true! Is there always a prime number between consecutive squares? No one knows.

It is our hope that this book will inspire some students to dedicate themselves to mathematics and/or computer science. Perhaps we can pass the torch to some young reader (though neither author intends to relinquish it just yet!). At any rate, enjoy this book, and please do the exercises.

The only prerequisites to understanding the material presented here are high school algebra, very basic programming skills, and a willingness to work. Read with pencil and paper handy. When in doubt, you should reread, recalculate, and rethink to your heart's content. In fact, scribble in the margins! In my opinion, that's what they are there for. A marginal comment written by the great seventeenth-century French mathematician Fermat sparked an exciting and productive 300-year long search for a proof that ended within the last decade. Fermat asserted, without proof, that while the sum of two squares is often a square ($16 + 9 = 25$), this is never true for cubes, fourth powers, fifth powers, etc. In other words, the equation $a^n + b^n = c^n$ has no solution in positive integers when $n > 2$.

Do also try the programming exercises. You can study the sample programs or try first on your own. It is like having a silent partner with perfect abilities at following directions.

Thanks to Anthony Delgado, Tim Bocchi, and Brian Phillips for their helpful suggestions.

Enjoy the adventure.

Marty Lewinter, PhD Mathematics
Jeanine Meyer, PhD Computer Science
June 2014

WORDS

In a secret chamber of my heart
Where there's no room for a lie,
Not even one small tiny part,
Where there's found no alibi,
In that closely guarded center
Behind thick walls that none can scale,
Where deceit may never enter
And hypocrisy will fail,
There's a truth that's carved in stone,
One that took some time to chisel—
Monolithic, cold, alone—
Yet it makes my spirit sizzle.

Often have I, silent, prayed
The solemn holy words unspoken,
Words that I have not betrayed—
And to this day my faith's unbroken.

Remember what you now shall read
Words you ne'er again might see.
Take them with you—pay them heed—
Reflect on them most carefully.

"Do not trust the words of others—
Parents, teachers, friends, or brothers.
Alone the wine of wisdom drink,
For truth comes but to those who think.

And when some people bully you,
Dictating what to think or do,
Fight them with the battle cry
That guards your mind with the sacred word: why?"

NOTATION IN MATHEMATICAL WRITING AND IN PROGRAMMING

The language of mathematics for expressions, equations, and inequalities is similar to what you see and what you produce for programming, but there are differences.

For example, if you see **ab** in this book or any other mathematics text, you would interpret it as **a** times **b**. To indicate the product of two variables in JavaScript (and most, maybe all, programming languages), you need to write a*b.

In programming, a variable is a construct for holding a value. If you see a+b in JavaScript, it could mean the arithmetic sum of the values stored in the variables a and b, but it also could mean concatenate the character string stored in a with the character string stored in b. This is true for some other programming languages, but not all. Concatenation of strings is not that frequent an occurrence in mathematical works on number theory, but it is used in some of the programming examples, mainly to produce something to display.

Another issue is that single character names may not be the best practice in programming. In the examples for this book, I often do use the names used in the text, but I also sometimes use longer, meaningful names, for my benefit and the benefit of the reader.

An equation in mathematics uses the equal sign = . You will see plenty of equal signs in code, but the meaning is different. For example,

```
a = b+c;
```

is an instruction to JavaScript to determine the values of variables b and c, add them (or concatenate), and then plug the result into variable a. One way to read it is "make a equal to $b + c$." With this in mind, the statement

```
a = a+1;
```

is reasonable. It is an instruction to determine the current value of the variable a, add 1 to it, and then plug in the result back into the variable a. One way to read it is "make a equal to the old value of a plus 1." This is a very common operation, and so there is a shorthand for it:

```
a++.
```

There is another shorthand for adding anything to a variable and resetting the variable with the new value. Here is an example using the concatenation of strings meaning of the + operator:

```
output+=String(val)+" ";
```

This means take the value of val and produce a string (if val had the value 5, then String(val) has the value "5") and add onto it a blank. (The effect of using   is to guarantee a blank value. It is required in certain situations in HTML.) The result of all these operations is then added onto the variable output. This is used in many of the programs for this book to display results.

In JavaScript, you will see a double equal sign, ==, often in if statements. The double equal sign is a comparison operator and yields a result of true or false. (True and false are values in JavaScript, called Boolean values, after George Boole.) For example,

```
if (c==-1) {
   deficient++; }
else if (c==1) {
   abundant++; }
else {
  perfect++; }
```

checks out the value of a variable c and, depending on whether it has the value -1, 1, or anything else, increments certain other variables. JavaScript has other comparison operations, for example > and >=. If you see a single equal sign in an if statement, it probably is a mistake.

Mathematics can make use of superscripts and subscripts, square root symbol, summation symbol, multiple line expressions, and other things. Programming code must make do with one line and a limited number of symbols (no Greek letters). As I indicated, we can and should use meaningful names for variables and functions, and we can use techniques such as under_scores and camelCasing.

To indicate 2^p, we make use of `Math.pow(2,p)`. For square root, we can use `Math.sqrt(n)`. Situations involving subscripts may be appropriate places to use arrays. To get to the kth element of an array myScores, the JavaScript is

```
myScores[k].
```

The index values (the indices) for arrays and for character strings start with 0, not 1.

Don't be worried if you don't understand all of this right now. It will become clear when you see the code in context and as you do your own programming. The important thing is to be ready to see similarities and differences between the languages of mathematics and programming.

There are other programming languages that have an even more mathematical nature than JavaScript. For example, check out the language Haskell. It supports infinite sets!

1

SPECIAL NUMBERS: TRIANGULAR, OBLONG, PERFECT, DEFICIENT, AND ABUNDANT

We start our introduction to number theory with definitions, properties, and relationships of several categories of numbers.

TRIANGULAR NUMBERS

Triangular numbers are those that can be written as the sum of a consecutive series of (whole) numbers beginning with 1. Thus 6 is triangular because it is the sum of the first three numbers: $6 = 1 + 2 + 3$. The first few triangular numbers are 1, 3, 6, 10, 15, 21, 28, 36, 45, and 55. We denote the *n*th triangular number by t_n. Thus $t_5 = 1 + 2 + 3 + 4 + 5 = 15$. More generally,

$$t_n = 1 + 2 + 3 + \cdots + (n-2) + (n-1) + n \qquad (1.1)$$

Our first program, calculating a specific triangular number, shows the format of an HTML document. The first line specifies the **doctype**. The rest is an *html* element, starting with **<html>** and ending with **</html>**. Within the html element is a *head* element and a *body*

Elementary Number Theory with Programming, First Edition. Marty Lewinter and Jeanine Meyer.
© 2016 John Wiley & Sons, Inc. Published 2016 by John Wiley & Sons, Inc.

element. In this case, the body element is empty. The head element contains a **meta** tag specifying the character type (it can be omitted), a title, and a *script* element. All the action is in the script element.

The code makes use of standard programming constructs such as variables and functions and for-loops (if you don't understand what these terms are, please consult any beginner book on programming. Shameless plug: go to *The Essential Guide to HTML5: Using Games to Learn HTML5 and JavaScript*, http://www.apress.com/9781430233831).

The specific triangular number we want is specified in the coding by setting the variable **n**. This is termed *hard-coding*. The computation is done using a for-loop. The for-loop adds up the values from 1 to *n*, exactly following Equation 1.1. The built-in method `document.write` writes out the result.

The challenge in Exercise 1 is to compare coding using Equation 1.1 versus Equation 1.2. The challenge is that computers are very fast. I use the built-in `Date` function with the method `getTime` to get the number of milliseconds from a base date at the start and after the computation. It turns out that computing the millionth triangular number takes 3 ms! You can experiment with different values. Using the formula given in Equation 1.2 would be much, much faster. Give it a try.

The *n*th triangular number is given by the formula:

$$t_n = 1 + 2 + 3 + \cdots + n = \frac{n(n+1)}{2} \qquad (1.2)$$

Example: $t_{100} = \dfrac{100 \times 101}{2} = 5{,}050$

Example: Write $6 + 7 + 8 + 9 + 10 + 11$ as the difference of two triangular numbers. We observe that $6 + 7 + 8 + 9 + 10 + 11 = (1 + 2 + 3 + 4 + 5 + 6 + 7 + 8 + 9 + 10 + 11) - (1 + 2 + 3 + 4 + 5)$, which is $t_{11} - t_5$.

Example: Generalize the previous example to any consecutive sum such as $45 + 46 + \cdots + 987$. Note that $a + (a+1) + (a+2) + \cdots + b = (1 + 2 + 3 + \cdots + b) - (1 + 2 + 3 + \cdots + (a-1)) = t_b - t_{a-1}$. By letting $a = 6$ and $b = 11$, we get the result of the previous example.

It should be noted that

$$t_n - t_{n-1} = n \tag{1.3}$$

The sum of any two consecutive triangular numbers is a square. For example, $t_4 + t_3 = 10 + 6 = 16 = 4^2$ and $t_5 + t_4 = 15 + 10 = 25 = 5^2$. This is expressed by the formula

$$t_n + t_{n-1} = n^2 \tag{1.4}$$

Example: Verify (1.4) for $n = 10$. We have $t_{10} + t_9 = 55 + 45 = 10^2$.

Example: Find two triangular numbers whose sum is 900. Since $900 = 30^2$, we have $n = 30$. Then using (1.4), $900 = t_{30} + t_{29} = \dfrac{30 \times 31}{2} + \dfrac{29 \times 30}{2} = 465 + 435$.

The sum of the reciprocals of all the triangular numbers is 2. Formally,

$$\frac{1}{1} + \frac{1}{3} + \frac{1}{6} + \frac{1}{10} + \cdots + \frac{1}{t_n} + \cdots = 2 \tag{1.5}$$

OBLONG NUMBERS AND SQUARES

A positive integer of the form $n(n + 1)$ is called *oblong*. The nth oblong number is the sum of the first n even numbers. To see this, observe that the nth even number is $2n$. Then we have $2 + 4 + 6 + \cdots + 2n = 2(1 + 2 + \cdots + n) = 2\left(\dfrac{n(n+1)}{2}\right) = n(n + 1)$, the nth oblong number. What about the sum of the first n odd numbers? The nth odd number is $2n - 1$. So $1 + 3 + 5 + \cdots + (2n-1) = (2 \times 1 - 1) + (2 \times 2 - 1) + (2 \times 3 - 1) + \cdots + (2n-1)$, in which -1 appears n times. We then get $2(1 + 2 + 3 + \cdots + n) - n = 2\left(\dfrac{n(n+1)}{2}\right) - n = n(n+1) - n = n^2 + n - n = n^2$. So the sum of the first n odd numbers is n^2.

Example: The sum of the first 5 odd numbers is 25. (Check this: $1 + 3 + 5 + 7 + 9 = 25$.) More impressively, the sum of the first 100 odd numbers is $100^2 = 10,000$.

The great French mathematician LaGrange (1736–1813) showed in the late eighteenth century that every positive number can be written as a sum of four or fewer squares. Thus, for example, $30 = 25 + 4 + 1$.

Number theorists are fond of numbers, such as 40, which are the sum of only two squares (e.g., $40 = 36 + 4$).

The Pythagoreans computed the sum of the first n powers of 2. Let

a. $S = 1 + 2 + 4 + \cdots + 2^{n-1}$. Then

b. $2S = 2 + 4 + \cdots + 2^{n-1} + 2^n$.

Now subtract Equation (a) from Equation (b), and we get $S = 2^n - 1$. We have, then, the following formula:

$$1 + 2 + 2^2 + 2^3 + \cdots + 2^{n-1} = 2^n - 1 \tag{1.6}$$

With a minor change in the proof of (1.6), we obtain an analogous formula for the sum of the first n powers of any base. Let $S = 1 + a + a^2 + a^3 + \cdots + a^{n-1}$. Then $aS = a + a^2 + a^3 + \cdots + a^{n-1} + a^n$. Subtract the first equation from the second, and we get $(a - 1)S = a^n - 1$. Upon division by $a - 1$, we obtain the following formula:

$$1 + a + a^2 + a^3 + \cdots + a^{n-1} = \frac{a^n - 1}{a - 1} \tag{1.7}$$

DEFICIENT, ABUNDANT, AND PERFECT NUMBERS

The Pythagoreans classified all numbers as *deficient, abundant,* or *perfect*. Given a number, find all of its *proper* factors, that is, all numbers that go into it (with the exclusion of the given number). The proper factors of 30, for example, are 1, 2, 3, 5, 6, 10, and 15.

Generating a list of the factors of a number is easy in JavaScript (and other programming languages), though it appears tedious to us. The modulo operation, %, determines the remainder. So if n is the number and f is a candidate factor, then

```
n % f
```

will produce the remainder of n divided by f. If this is 0, then f is a factor. If f < n, then f is a proper factor.

> The program uses a for-loop going from 1 up to but not including n. If it is a factor, the number is written out in the html document using `document.write` and a variable `count` is incremented.

If the sum of the proper factors of n is less than n, we call n *deficient*. If the sum exceeds n, it is called *abundant*. If the sum equals n, we call it *perfect*. For example, 8 is deficient since $1+2+4 < 8$, 18 is abundant since $1+2+3+6+9 > 18$, and 28 is perfect since $1+2+4+7+14 = 28$. The smallest perfect number is 6. The first few perfect numbers are 6, 28, 496, and 8128. It is not known today whether there are infinitely many perfect numbers. Moreover, all known perfect numbers are even. No one knows if there are any odd perfect numbers! Incidentally, the smallest abundant odd number is 945, while the smallest abundant even number is 12.

> The program to characterize a number as deficient, perfect, or abundant was made by modifying the previous one that listed and counted the number of proper factors. To make the determination of whether a number n is deficient, perfect, or abundant, the program has to add up the proper factors. So the statement count++ is removed and the statement
>
> ```
> sum += i;
> ```
>
> is inserted. By the way, this is shorthand for taking the original value of the variable `sum`, adding one to it and then assigning that back to the variable `sum`.
>
> ```
> sum = sum + i;
> ```
>
> I also changed the name of the function to `addUpFactors`. I tested the program using the specific numbers given in the text.

The Pythagoreans found an amazing method for finding perfect numbers. They observed, using (1.6), that sums of the form $1+2+2^2+2^3+\cdots+2^{n-1}$ are *prime* for certain values of n and are *composite* for others. (A number is *prime* if its only factors are 1 and itself. 7, 19, and 31 are examples of primes. A *composite* number has proper factors other than 1. Thus 20 is composite.) The following sums, for example, are prime:

$$1 + 2 = 3$$

$$1 + 2 + 4 = 7$$

$$1 + 2 + 4 + 8 + 16 = 31$$

$$1 + 2 + 4 + 8 + 16 + 32 + 64 = 127$$

In each of these equations, multiply the greatest number on the left by the number on the right, yielding $2 \times 3 = 6$, $4 \times 7 = 28$, $16 \times 31 = 496$, and $64 \times 127 = 8128$. These products, 6, 28, 496, and 8128 are perfect. Whenever the sum of the first n powers of 2 is prime, this procedure yields a perfect number! Using (1.6), the sum of the first n powers of 2 is $2^n - 1$, so the perfect number is of the form $2^{n-1}(2^n - 1)$. The prime sum, $2^n - 1$, is then called a *Mersenne prime*, in honor of the eighteenth century French mathematician. It was shown in the eighteenth century by the great Swiss mathematician Leonhard Euler (1707–1783) that all even perfect numbers are of the form $2^{n-1}(2^n - 1)$.

A conjecture is that no odd number (odd number >1) is perfect. One of the exercises and one of our programs tests this conjecture on the first 1000 odd numbers.

The program to test the conjecture concerning odd numbers not being perfect numbers is built on the previous example. Instead of writing out the result, a function with a parameter is made to return -1, 0, or 1 if the number is deficient, perfect, or abundant. Any three distinct values could be used.

The inner function (I named it `classify`) is called for all odd numbers up to a limit. The task is to determine how to generate the set of numbers. A solution is to use a for-loop going from `j=1 to 1000` and, within the loop, setting a variable n to `2*j+1`.

It is very important to keep in mind that this is not a proof of the conjecture. Something could be happening at higher numbers.

The Pythagoreans believed that if two friends wore amulets, one with 220 and the other with 284, they would fortify their friendship. This is because the sum of the proper factors of either one of these numbers equals the other number, that is,

$$220 = 1 + 2 + 4 + 71 + 142$$

$$284 = 1 + 2 + 4 + 5 + 10 + 11 + 20 + 22 + 44 + 55 + 110$$

We call a pair of numbers with this property, *amicable numbers*. 1184 and 1210 comprise the next pair of amicable numbers, since

$$1210 = 1 + 2 + 4 + 8 + 16 + 32 + 37 + 74 + 148 + 296 + 592$$
$$1184 = 1 + 2 + 5 + 10 + 11 + 22 + 55 + 110 + 121 + 242 + 605$$

EXERCISES

An asterisk (∗) indicates that the exercise can be developed into a research project.

1 Write a program to find the nth triangular number, t_n, using formula (1.1). Then write a program using (1.2). Compare the two procedures for very large values of n. *A program for the first part of this is included at the end of the chapter.*

2 Write a program to find out whether a given number is a square.

3 Find, using a partial fraction decomposition, the sum of the reciprocals of the first n triangular numbers, that is, find $\frac{1}{1} + \frac{1}{3} + \frac{1}{6} + \frac{1}{10} + \cdots + \frac{1}{t_n}$. Then write a program to do this.

4 ∗The ancient Egyptians expressed every proper fraction except as the sum of fractions with 1s in the numerator. Thus, $\frac{2}{3}$ is equivalent to $\frac{1}{3}$ $+ \frac{1}{4} + \frac{1}{12}$ and $\frac{7}{8}$ is equivalent to as $\frac{1}{2} + \frac{1}{4} + \frac{1}{8}$.

   a. Verify the identity $\frac{1}{n} = \frac{1}{(n+1)} + \frac{1}{(n(n+1))}$.

   b. Using the strategy of starting off by writing a fraction a/b as the sum of 1/b + 1/b + ⋯ (a times) and using the identity verified in part a repeatedly until all fractions are distinct, write a program to express every fraction as the sum of fractions with distinct denominators and numerators equal to 1.

5 ∗Note that the sum and difference of the triangular numbers 15 and 21 are triangular. Verify that this is also the case for the triangular numbers 780 and 990. Find 100 more cases.

6 Find 100 triangular numbers that are squares.

7 From the first 1000 triangular number, find ones that are the sum of two other triangular numbers.

8   *Find 100 oblong numbers that are products of an oblong number and a square.

9   Show that after 3, the next 100 triangular numbers are composite (not prime). Then prove this for all triangular numbers after 3.

10 Write each of the numbers from 1 to 1000 as the sum of three or fewer triangular numbers.

11 *Write each of the numbers from 1 to 1000 as the sum of four or fewer squares. For which of these numbers can this be done in more than one way? For example, $50 = 49 + 1 = 25 + 25 = 36 + 9 + 4 + 1 = 16 + 16 + 9 + 9$.

12 *What proportion of the first 1000 numbers can be written using two or fewer squares?

13 Write a program that lists the proper divisors of a given number. *A program for this is given at the end of the chapter.*

14 Write a program to find the sum of the proper divisors of a given number.

15 Modify the program of the previous exercise to decide whether a given number is deficient, perfect, or abundant. *A program for this is included at the end of the chapter. You can improve this in various ways, including having the user enter the number. Look ahead to an example in Chapter 2 that shows how to get user input.*

16 *Write a program to check for perfect numbers within a range. You can set the endpoints of the range within the program. You can research to find a list of the known perfect numbers AND to determine the biggest integer value that can be represented in regular JavaScript.

17 *A number is called *semi-perfect* if it is the sum of *some* (but not all) of its proper divisors. 12 is the smallest semi-perfect number since $12 = 6 + 4 + 2$. Find the next 50 semi-perfect numbers.

18 *If a given number is abundant, determine if it is semi-perfect.

19 Show that 2^n is deficient for all $n \le 25$. Then show it is deficient for all n.

20 *Verify that 945 is the smallest odd abundant number. Find the next 10 odd abundant numbers. Do they seem to be getting further apart?

21 Observe that the square of the triangular number 6 is also triangular. Verify that this does not occur for any other triangular number (except 1) up to t_{1000}.

22 It has been conjectured that no odd number is perfect. Verify this for the first 1000 odd numbers. *See last example. You can decide on other ways to present the findings and also change the limit.*

Triangular Numbers

```
<html>
<head>
<title>Triangular Numbers</title>
<script>
var n = 1000000;
var start = new Date();
start = start.getTime();

function init(){
    var sum = 0;
  for(i=1;i<=n;i++){
    sum+=i;
  }
  now = new Date();
  now = now.getTime();
  elapsed = (now - start);
  document.write("The "+n+
      "th triangular number is "+sum+".<br/>");
  document.write("Elapsed time was "+elapsed+"
  milliseconds.");
}
init();
</script>
</head>
<body>
</body>
</html>
```

Proper Factors

```
<!DOCTYPE HTML>
<html>
<head>
<title>Proper factors</title>
<script>
var n = 30;
```

```
var count = 0;
function countUpFactors(n){
  document.write("Proper factors of "+n+" are:
  <br/>");
  for (var i=1;i<n;i++){
      if ((n%i)==0){
          count++;
          document.write(i+"<br/>");
      }
  }
  document.write("The number of proper factors
  of "+n+" is "+count+".");
}
countUpFactors(n);
</script>
</head>
<body>
</body>
</html>
```

Classifying Number as Deficient, Perfect, or Abundant

```
<!DOCTYPE HTML>
<html>
<head>
<title>Perfect or </title>
<script>
function addUpFactors(n){
  document.write("Proper factors of "+n+" are:
  <br/>");
  var sum = 0;
  for (var i=1;i<n;i++){
    if ((n%i)==0){
        document.write(i+"<br/>");
        sum += i;
    }
  }

  document.write("The sum of the proper factors
  of "+n+" is "+sum+", so "+n+" is ");
  if (sum<n){
```

```
      document.write("deficient.");
  }
  else if (sum==n) {
      document.write("perfect.");
  }
  else {
      document.write("abundant.");
  }
  document.write("<br/>");
}
addUpFactors(6);
addUpFactors(8);
addUpFactors(18);
addUpFactors(28);
addUpFactors(945);
addUpFactors(8128);
</script>
</head>
<body>
</body>
</html>
```

Checking the Conjecture on No Perfect Odd Number (up to 1000)

```
<!DOCTYPE HTML>
<html>
<head>
<title>Sort Odd numbers </title>
<script>
function classify(n){

   var sum = 0;
   for (var i=1;i<n;i++){
       if ((n%i)==0){
            //document.write(i+"<br/>");

            sum += i;
       }
   }

   if (sum<n) {
       return -1;
```

```
    }
    else if (sum==n) {
        return 0;
    }
    else {
        return 1;
    }
    document.write("<br/>");
}
 function sortOdds(limit) {
 var perfect = 0;
 var deficient = 0;
 var abundant = 0;
 for (var j=1;j<=limit;j++) {
     var n=2*j+1;
     c = classify(n);
     if (c==-1){
      deficient++;
     }
     else if (c==1) {
      abundant++;
     }
     else {
      perfect++;
     }
 }
 document.write("First "+limit+" odd numbers:
 <br/>");
 document.write ("deficient: "+deficient+
 "<br/>");
 document.write("abundant: "+abundant+
 "<br/>");
 document.write("perfect: "+perfect+
 "<br/>");
 }
 sortOdds(1000);
</script>
</head>
<body>
</body>
</html>
```

2

FIBONACCI SEQUENCE, PRIMES, AND THE PELL EQUATION

A prime number is a number that has no divisors other than itself or 1.
The subject of prime numbers is a big part of number theory.

PRIME NUMBERS AND PROOF BY CONTRADICTION

The first few primes are 2, 3, 5, 7, 11, 13, 17, and 19. A number that has *divisors* (or *factors*) other than itself and 1 is called *composite*. A composite number must have a prime divisor. Let *n* be a composite number. Then it has a factor, say, *m*, smaller than *n*. For example, 100 has the divisor 25. Now, if *m* is a prime, we have the desired prime divisor. If, on the other hand, *m* is composite, then *m* has a smaller divisor, say, *k*. Once again, if *k* is a prime, we have the desired prime divisor. If not, continue the process by producing a still smaller divisor, say, *j*, of *k*. Note that these divisors (*m*, *j*, and *k*) are divisors of the original number, *n*. Now, this process can't go on forever, since *n* has finitely many divisors. It follows that sooner or later, this process produces a prime divisor.

Elementary Number Theory with Programming, First Edition. Marty Lewinter and Jeanine Meyer.
© 2016 John Wiley & Sons, Inc. Published 2016 by John Wiley & Sons, Inc.

The great Greek geometer and number theorist Euclid, who lived in Alexandria, Egypt, circa 300 b.c., proved that there are infinitely many primes. He began by assuming that there are only finitely many primes and proceeded to obtain a *contradiction*. (The proof may have been known before Euclid wrote it down in his 13-volume book, "Elements.")

If there are finitely many primes, call them a, b, c, and so on up to the supposed last (or greatest) prime, say, p. Then any number larger than p is a composite number, since we are assuming that p is the largest prime. Now, let $N = abc...p$ (i.e., N is the product of all the primes), and then consider the larger number $N + 1$ (same as $abc...p + 1$). Since this number is greater than p, it is a *composite* number. Then it must have a *prime* divisor. But this prime divisor is clearly a divisor of N (since N is the product of all the primes) and cannot, therefore, be a factor of $N + 1$. This is a contradiction! It follows that the initial assumption that there is a largest prime is mistaken. Then there are infinitely many prime numbers.

Example: Show using proof by contradiction that if the composite number n equals ab, where $1 < a \le b < n$, then $a \le \sqrt{n}$. Assume to the contrary that $a > \sqrt{n}$. Then since $b \ge a$, we have $b \ge \sqrt{n}$. Then $ab > \sqrt{n} \times \sqrt{n}$, in which case $ab > n$, a contradiction.

As a consequence of the above example, to determine whether a given number is a prime, it suffices to check whether it is divisible by any of the numbers less than or equal to its square root.

Example: Show that 41 is a prime number. Since $6 < \sqrt{41} < 7$, we only need to check the possible factors 2, 3, 4, 5, and 6.

Let's write a program to produce a list of all the primes up to 1000.

We construct a program with a for-loop going from 3 to 1000. At each step, we check if there are any factors. If a number has a factor, the number is not a prime. That is, our checking tests for factors. If no factors appear, then the number is a prime.

Our code will have a for-loop for candidate primes going from 3 to our limit (say, 1000):

```
for (n=3;n<=1000;n++)
```

The next step is to test for factors. You have read that we only need to test up to the square root of the number. We use a built-in function for square root:

```
s = Math.sqrt(n);
```

We have a loop within the loop that produces the candidates for factor:

```
for (f=2;f<=s;f++)
```

Keep in mind that the square root may not be an integer. The for statement produces loop variables, the f's, that *are* all integers and compares with the value s.

How to check if a number f is a factor of a number n? Recall from the last chapter, if

```
r = n % f;
```

The variable r will hold the remainder. If r is equal to zero, then f is a factor of n.

Now, for our task, we just need one factor, so if this is true (r is equal to zero), we (our code) are done with checking n. It is not a prime. Our code moves on to the next value of n. The technique for this operation uses what is termed a flag variable. My name captures what it is: noFactorSoFar. It is set equal to true before the inner loop checking for factors. As soon as we find a factor, our code sets noFactorSoFar to false. In the program, the code checks noFactorSoFar. If it is still true, then n is a prime; otherwise, it is not a prime.

When you look at the program, you see a loop within a loop and note that a whole lot of stuff is going on, lots of modulo operations. The leap you need to make, and it is as much psychological as mathematical, is to realize that you are not doing the work, the computer is. The code is at the end of the chapter. Also, be aware that you can view the source code of any HTML/JavaScript document on the browser. Look over this example and then come right back.

Example of 41 continued: Since 2 does not go into 41, neither does 4 or 6. Finally, since neither 3 nor 5 goes into 41, we are done. 41 is a prime number.

Using the idea that we don't have to check for division by 4 or 6 if 2 is not a divisor, we can improve the program! Yes, the program is doing all the work, but…let's try to be more efficient. Instead of checking *all* numbers up to the square root as candidate factors, let's just check primes. If the number *n* is divisible by something, it is divisible by a prime. But you now may ask: how do we get a list of primes? Isn't that the point of the exercise? If we had a list, why would we need to check? The answer is: yes, of course, at the start there isn't a list, but we can make a list as we go along. Technically, what we do (the code does) is start off with an array holding exactly one number, the number 2. (Note: 2 is sometimes not considered a prime and sometimes is…. We need it on our list.) Whenever we determine that something is a prime, we add it to the list. This is done using a built-in JavaScript method called `push`. So at the start, we will have

```
var primes = [2];
```

and in our code, when it is determined that a specific *n* is a prime, we will have

```
primes.push(n);
```

The header for the inner loop will be slightly more complex. The loop variable, shown here with the variable name `fi`, serves as the index into the `primes` array. The stopping condition has two parts: checking the length of the array (remember: the index values for arrays go from 0 to 1 less than the length of the array) and *then* checking if the factor, the particular prime in the array, is less than or equal to the square root of *n*. The header is

```
for (fi=0;fi<primes.length && primes[fi]<=s;fi++)
```

This compound logical test works because the program, as I indicated when I used the word *then*, does the first test, `fi < primes.length`, and only does the second if the first is true. This attribute of the `&&` operator is called "lazy" and it is an appropriate term.

> The code in the `fi` loop immediately sets the variable `f` to be the element in the `primes` array at index `fi`:
>
> ```
> f = primes[fi];
> ```
>
> The code continues as in our first example.
>
> The next step is that we put in some bookkeeping, also known as instrumentation, to see if checking only primes was a gain. The thing to measure is the number of mod operations. I also put in the number of primes found. Look at both examples and run the programs. The results are that there is a big improvement in performance: only 2800 mod operations compared to 5287 when the limit is 1000.
>
> There are three programs: the simplest one, checking all numbers as candidate factors; a version of this program with the bookkeeping code, that is, counting the operations; and the improved version, also with the bookkeeping.

All numbers have exactly one of the forms $4k$, $4k+1$, $4k+2$, and $4k+3$. Thus, 7, for example, is of the form $4k+3$, as can be seen by letting $k=1$. On the other hand, 21 is of the form $4k+1$, as can be seen by letting $k=5$. It is obvious that numbers of the form $4k$ or $4k+2$ are composite (except for 2, the only even prime). This leaves the two "classes" of numbers of the forms $4k+1$ and $4k+3$. Each class contains infinitely many primes.

PROOF BY CONSTRUCTION

Many proofs employ *construction*, that is, the claim of the theorem is proven by constructing a general example. Here are two theorems illustrating this idea. To understand the first proof, recall that $n!$, read "n factorial," is the product of the first n positive integers, that is,

$$n! = n \times (n-1) \times (n-2) \times \cdots \times 3 \times 2 \times 1$$

Theorem: There are arbitrarily lengthy sequences of consecutive composite numbers.

Proof: Consider the sequence of $n-1$ consecutive numbers $n!+2, n!+3, n!+4, \ldots, n!+n$. These are all composite. To see this, observe that 2 is a factor of $n!+2$, 3 is a factor of $n!+3$, 4 is a factor of $n!+4, \ldots$, and finally,

n is a factor of $n! + n$, and the proof is over, since n may be chosen as large as we please.   ☐

SUMS OF TWO SQUARES

Theorem: Let x and y be integers that can be written as the sum of two squares. Then their product, xy, can also be written as the sum of two squares.

Proof: Let $x = a^2 + b^2$ and $y = c^2 + d^2$. Then it is easy to verify that $xy = (a^2 + b^2)(c^2 + d^2) = (ac + bd)^2 + (ad - bc)^2$.   ☐

Remark: Another way to write xy as the sum of two squares (if $x \neq y$) is given by $xy = (ac - bd)^2 + (ad + bc)^2$.

Example: Since $50 = 5 \times 10$, and since $5 = 4 + 1$ and $10 = 9 + 1$, it follows by the above remark that 50 can be written as the sum of two squares in two different ways. Guesswork yields $25 + 25$ and $49 + 1$.

BUILDING A PROOF ON PRIOR ASSERTIONS

Many assertions in mathematics are proven by invoking prior proven assertions. Here is an example. Let us prove that n^k, where n and k are given positive integers greater than 1, is the sum of n consecutive odd numbers beginning with $n^{k-1} - n + 1$. That is, show that

$$n^k = \left(n^{k-1} - n + 1\right) + \left(n^{k-1} - n + 3\right) + \left(n^{k-1} - n + 5\right) + \cdots$$
$$+ \left(n^{k-1} - n + 2n - 1\right)$$

Note that the constants in parentheses are the first n odd numbers, 1, 3, 5, ..., $2n - 1$, which we know have the sum n^2. Note also that the expression $n^{k-1} - n$ occurs in each of the n parentheses. Then the sum on the right in the above equation is $n(n^{k-1} - n) + n^2 = n^k - n^2 + n^2 = n^k$, and the proof is complete.   ☐

Example: When $n = 3$ and $k = 4$, we get $3^4 = 25 + 27 + 29 = 81$.

When $k = 3$, we get $n^3 = (n^2 - n + 1) + (n^2 - n + 3) + (n^2 - n + 5) + \cdots + (n^2 - n + 2n - 1)$, proving the following theorem.

Theorem: n^3 is the sum of n consecutive odd numbers starting with $n^2 - n + 1$.   □

SIGMA NOTATION

Sigma notation is a convenient device for writing large sums. It involves a variable that is present in each term and that changes values consecutively. Thus, the sum $1 + 2 + 3 + \cdots + 100$ can be written by employing the variable m and noting that m goes from 1 to 100, consecutively. The command to add the terms is simply $\sum_{m=1}^{100} m$. The numbers under and over the capital sigma indicate where to start and where to stop. The function of m in front of the sigma indicates the typical expression being added.

Example: To add the first 10 squares, that is, to denote $1 + 4 + 9 + \cdots + 100$, we write $\sum_{m=1}^{10} m^2$.

Example: To add the first 10 cubes, that is, to denote $1 + 8 + 27 + \cdots + 1000$, we write $\sum_{m=1}^{10} m^3$.

Now, some sums are tricky in that the number of terms is variable, such as "the sum of the first n odd numbers." Since the mth odd number is $2m - 1$, we write $\sum_{m=1}^{n} 2m - 1$. We call m a *dummy variable* since, unlike n, it doesn't survive the addition, and it varies as we evaluate the expression. (Note that n stays constant.) Note that many sums employing sigma can be computed with a simple "for-loop."

SOME SUMS

Here are some formulas for the sum of the first n powers:

$$\sum_{m=1}^{n} m^2 = 1 + 4 + 9 + \cdots + n^2 = \frac{n(n+1)(2n+1)}{6} \tag{2.1}$$

$$\sum_{m=1}^{n} m^3 = 1 + 8 + 27 + \cdots + n^3 = (1 + 2 + 3 + \cdots + n)^2 = \left[\frac{n(n+1)}{2}\right]^2 \tag{2.2}$$

Recall that an *oblong* number has the form $k(k+1)$. The first few oblong numbers are 2, 6, 12, 20, and 30. The sum of the first n oblong numbers is given by

$$\sum_{m=1}^{n} m(m+1) = 2 + 6 + 12 + \cdots + n(n+1) = \frac{n(n+1)(n+2)}{3} \quad (2.3)$$

If we divide both sides of (2.3) by 2, we get, as a bonus, the following formula for the sum of the first n triangular numbers:

$$\sum_{m=1}^{n} t_m = \sum_{m=1}^{n} \frac{m(m+1)}{2} = 1 + 3 + 6 + \cdots + \frac{n(n+1)}{2} = \frac{n(n+1)(n+2)}{6}$$

$$(2.4)$$

FINDING ARITHMETIC FUNCTIONS

Let n be an integer. Then $f(n)$ is *arithmetic* if $f(n)$ is an integer. Thus, $f(n) = 2n^3$ is arithmetic. We present a way of finding arithmetic functions from several values of $f(n)$. Look at the chart plotting n against $f(n)$, for $n = 1, 2, 3$, and 4. Let us find a function that does this.

n	$f(n)$
1	5
2	8
3	11
4	14

Infinitely many functions can do the job. But if we assume that the pattern of values persists ($f(n)$ seems to grow by the same amount, viz., 3, from row to row, i.e., the so-called first differences, $f(n+1) - f(n)$, are constant), the function might be *linear*, or *first degree*. Then we set $f(n) = an + b$ and proceed to find a and b. Two unknowns generally require two equations, so we use the facts that $f(1) = 5$ and $f(2) = 8$. This yields the simultaneous equations:

$$a + b = 5$$
$$2a + b = 8$$

with solution $a = 3$ and $b = 2$. Our function is $f(n) = 3n + 2$. This works if the first differences of the function f are constant. The first differences are constant if and only if $f(n)$ is of the form $an + b$. A calculation of the first differences yields $f(n + 1) - f(n) = a(n + 1) + b - (an + b) = an + a + b - an - b = a$, which is a constant, independent of n. If you plot the points in the chart, they will lie on a straight line, that is, the line $y = ax + b$. Now, a straight line has constant slope, namely, $\dfrac{\Delta y}{\Delta x}$. Since the x values are 1, 2, 3, ..., we have $\Delta x = 1$. Since the y values jump by the constant value a, we have $\Delta y = a$. Then the slope is a, which is in perfect agreement with the equation $y = ax + b$.

Assuming, from the handful of known values for the first few integers, that a function is a polynomial, how do we infer its degree from its chart of values and how do we determine the coefficients? To answer these questions, let's define *kth order differences*. The next chart shows $f(n)$ and its successive differences $\Delta_1 f$, $\Delta_2 f$, and $\Delta_3 f$. Now, $\Delta_1 f$ is the first difference we used in the previous problem.

n	$f(n)$	$\Delta_1 f$	$\Delta_2 f$	$\Delta_3 f$
1	3	8	12	6
2	11	20	18	6
3	31	38	24	6
4	69	62	30	6
5	131	92	36	
6	223	128		
7	351			

The fourth column, labeled $\Delta_2 f$, yields the differences of $\Delta_1 f$, which are called the *second differences* of $f(n)$. Similarly, the quantities $\Delta_3 f$ are called the *third differences*. This process could go on forever, but if the arithmetic function $f(n)$ is a polynomial of degree k, it can be shown that $\Delta_k f$ is constant so all differences of order greater than k are zero.

Once it is determined from a chart (again, assuming persistence of the pattern) that $f(n)$ has degree k, one can find the $k + 1$ coefficients of the polynomial using the first $k + 1$ values of f. Since the third differences are constant here, we require a cubic function, $f(n) = an^3 + bn^2 + cn + d$. The determination of the four coefficients requires four equations, in turn requiring the first four values $f(1) = 3$, $f(2) = 11$, $f(3) = 31$, and $f(4) = 69$. We get four equations in four unknowns by letting $n = 1$, 2, 3, and 4 in the equation for $f(n)$:

$$a + b + c + d = 3$$
$$8a + 4b + 2c + d = 11$$
$$27a + 9b + 3c + d = 31$$
$$64a + 16b + 4c + d = 69$$

whose solution $a = 1$, $b = 0$, $c = 1$, and $d = 1$ can be obtained by first subtracting each of first three equations from its successor, yielding three equations in three unknowns:

$$7a + 3b + c = 8$$
$$19a + 5b + c = 20$$
$$37a + 7b + c = 38$$

Repeating the same technique yields

$$12a + 2b = 12$$
$$18a + 2b = 18$$

whose solution is $a = 1$, $b = 0$. Plugging these values into the original four equations in four unknowns yields two equations in two unknowns. We finally obtain the desired function, $f(n) = n^3 + n + 1$. Note that this is true for every value listed in the chart.

Example: Let $f(n)$ be the sum of the integers from 1 to n. Putting the first few values of $f(n)$ into a chart gives us:

n	$f(n)$	$\Delta_1 f$	$\Delta_2 f$
1	1	2	1
2	3	3	1
3	6	4	1
4	10	5	
5	15		

Since the second differences are constant, we look for a quadratic function, $f(n) = an^2 + bn + c$. Using the first three values $f(1) = 1$, $f(2) = 3$, and $f(3) = 6$, we get the three equations

$$a + b + c = 1$$
$$4a + 2b + c = 3$$
$$9a + 3b + c = 6$$

yielding $a = \dfrac{1}{2}$, $b = \dfrac{1}{2}$, and $c = 0$. Then $f(n) = \dfrac{1}{2}n^2 + \dfrac{1}{2}n = \dfrac{n(n+1)}{2}$.

FIBONACCI NUMBERS

In the sequence 1, 1, 2, 3, 5, 8, 13, 21, 34, 55, 89, 144, …, each term is the sum of the preceding two terms. We denote these *Fibonacci numbers* by

u_1, u_2, u_3, ... Thus, u_n is the nth Fibonacci number. Now, $u_1 = 1$ and $u_2 = 1$ are the *seed* since they determine the rest of the terms. (For example, $u_3 = u_1 + u_2 = 1 + 1 = 2$.) The relationship between the general term and the two preceding terms is called the *recursive relation* and is written as

$$u_{n+2} = u_n + u_{n+1} \qquad (2.5)$$

When $n = 3$, this becomes $u_5 = u_3 + u_4 = 2 + 3 = 5$. While the Fibonacci numbers get arbitrarily large, an odd thing happens when we study the sequence of ratios of the form $\dfrac{u_{n+1}}{u_n}$, that is, the ratios consisting of each term divided by its predecessor. The first few such ratios are $\dfrac{1}{1} = 1$, $\dfrac{2}{1} = 2$, $\dfrac{3}{2} = 1.5$, $\dfrac{5}{3} = 1.66...$, $\dfrac{8}{5} = 1.6$, $\dfrac{13}{8} = 1.625$, which makes us suspect that the ratios approach a limit. The limit is the famous irrational number, denoted by the Greek letter ϕ, known as the *golden ratio*. (Its approximate value is 1.618.) To show this, divide both sides of (2.5) by u_{n+1}, yielding the equation

$$\frac{u_{n+2}}{u_{n+1}} = \frac{u_n}{u_{n+1}} + 1 \qquad (2.6)$$

Let L denote the limit of the ratios we seek. As n approaches infinity, the left side of (2.6) approaches L. (u_{n+1} is the predecessor of u_{n+2}.) On the other hand, the first term on the right side of (2.6) is the reciprocal of the ratio we are considering, since the *numerator* u_n is the predecessor of the *denominator* u_{n+1}. Then this term approaches $\dfrac{1}{L}$.

Inserting these values into (2.6) yields $L = \dfrac{1}{L} + 1$. Multiplying both sides by L and transposing everything to the left yields $L^2 - L - 1 = 0$, whose positive root yields L, or ϕ as it is commonly called. The positive root is $\dfrac{1 + \sqrt{5}}{2}$.

Suppose you live in a country whose currency consists solely of one and two dollar bills. Given n, we wish to count the number of ways of putting n dollars into a vending machine using one and two dollar bills. Let us agree to consider *order* as a differentiating factor, as well as the quantities of both kinds of bills. In technical terms, we are counting *ordered partitions* of n using 1's and 2's. Denote the function that counts these partitions by $f(n)$. When $n = 3$, for example, we have the following three possibilities:

$$2 + 1$$
$$1 + 2$$
$$1 + 1 + 1,$$

so $f(3) = 3$. When $n = 4$, we have the following five possibilities:

$$2 + 2$$
$$2 + 1 + 1$$
$$1 + 2 + 1$$
$$1 + 1 + 2$$
$$1 + 1 + 1 + 1,$$

indicating that $f(4) = 5$. Finally, when $n = 5$, we have the following eight possibilities:

$$2 + 2 + 1$$
$$2 + 1 + 2$$
$$1 + 2 + 2$$
$$2 + 1 + 1 + 1$$
$$1 + 2 + 1 + 1$$
$$1 + 1 + 2 + 1$$
$$1 + 1 + 1 + 2$$
$$1 + 1 + 1 + 1 + 1,$$

in which case, $f(5) = 8$. The cases $n = 1$ and $n = 2$ are easy. When $n = 1$, we have the single possibility, 1, while when $n = 2$, we have two possibilities, 2 and $1 + 1$, so $f(1) = 1$ and $f(2) = 2$.

In summary, the first five values of $f(n)$ are 1, 2, 3, 5, and 8. It seems like we obtained the Fibonacci numbers u_2, u_3, u_4, u_5, and u_6. One would be tempted to guess that $f(n) = u_{n+1}$. To prove this amazing fact, observe that it works in the first few instances. It suffices, therefore, to show that the function $f(n)$ obeys the recursive relation of the Fibonacci numbers, that is, we must show that $f(n + 2) = f(n) + f(n + 1)$.

Ordered partitions of $n + 2$ consisting of 1's and 2's can be divided into two piles. The first pile will include all partitions that begin with 2, while the second pile will consist of the remaining partitions, that is, those that begin with 1. The number of partitions in the first pile is clearly $f(n)$ since 2 is the first term of each of these partitions, requiring that we complete the process by partitioning n. By the same token, the number of partitions in the second pile is clearly $f(n + 1)$ since 1 is the first term of each of these partitions, requiring that we complete the process by partitioning $n + 1$. Thus, $f(n + 2) = f(n) + f(n + 1)$. Partitions play a significant role in number theory.

Here is a theorem that yields the sum of the first n terms of the Fibonacci sequence.

Theorem: The first n terms of the Fibonacci sequence satisfy

$$u_1 + u_2 + u_3 + \cdots + u_n = u_{n+2} - 1 \tag{2.7}$$

Proof: The recursive relation of the Fibonacci numbers can be rewritten as

$$u_k = u_{k+2} - u_{k+1}$$

Writing this for $k = 1, 2, 3, \ldots, n$ yields the following n equations:

$$u_1 = u_3 - u_2$$
$$u_2 = u_4 - u_3$$
$$u_3 = u_5 - u_4$$
$$\vdots$$
$$u_{n-1} = u_{n+1} - u_n$$
$$u_n = u_{n+2} - u_{n+1}$$

Now, adding the left and right sides of these equations and canceling on the right side yields

$$u_1 + u_2 + u_3 + \cdots + u_n = u_{n+2} - u_2 = u_{n+2} - 1,$$

which is exactly (2.7). $\qquad\square$

Theorem: The first n terms of the Fibonacci sequence satisfy the identity

$$u_1^2 + u_2^2 + u_3^2 + \cdots + u_n^2 = u_n u_{n+1} \tag{2.8}$$

There is a formula in terms of n for finding u_n. It is called *Binet's formula*.

$$u_n = \frac{1}{\sqrt{5}} \left[\left(\frac{1 + \sqrt{5}}{2} \right)^n - \left(\frac{1 - \sqrt{5}}{2} \right)^n \right] \tag{2.9}$$

Who would believe that a formula for the Fibonacci numbers could involve $\sqrt{5}$?

AN INFINITE PRODUCT

Consider the product of the expressions $1 - \dfrac{1}{k^2}$ as k goes from 2 to some given number, n. When $n = 5$, for example, we get

$$\left(1-\frac{1}{4}\right) \times \left(1-\frac{1}{9}\right) \times \left(1-\frac{1}{16}\right) \times \left(1-\frac{1}{25}\right)$$

$$= \left(\frac{3}{4}\right) \times \left(\frac{8}{9}\right) \times \left(\frac{15}{16}\right) \times \left(\frac{24}{25}\right) = \frac{3}{5}$$

Let us find the general product as a function of the given number, n. Observing that

$$1-\frac{1}{k^2} = \frac{k^2-1}{k^2} = \frac{(k-1)(k+1)}{(k)(k)},$$

we get

$$\left(1-\frac{1}{2^2}\right)\left(1-\frac{1}{3^2}\right) \cdots \left(1-\frac{1}{n^2}\right)$$

$$= \frac{1\times 3}{2\times 2} \cdot \frac{2\times 4}{3\times 3} \cdot \frac{3\times 5}{4\times 4} \cdot \frac{4\times 6}{5\times 5} \cdots \frac{(n-1)\times(n+1)}{n\times n} = \frac{1}{2} \cdot \frac{n+1}{n}$$

since everything else cancels. This last expression equals $\dfrac{n+1}{2n}$, or $\dfrac{1}{2} + \dfrac{1}{2n}$. (When $n = 5$, we get 3/5, which is in perfect agreement with the answer we obtained earlier.) Now, what happens if we let n got to infinity? In other words, what if we take the product of the expressions $1 - 1/k^2$ where k ranges over *all* the numbers from 2 to infinity? Since the product up to n is $\dfrac{1}{2} + \dfrac{1}{2n}$, and since the second fraction, $\dfrac{1}{2n}$, goes to zero as n goes to infinity, the answer is $\dfrac{1}{2}$.

THE PELL EQUATION

When are x^2 and $2y^2$ consecutive? $x = y = 1$ yields a solution since 1 and 2 are consecutive. The solution $x = 3$ and $y = 2$ produces the consecutive numbers 8 and 9. Indeed, 8 is twice a square and 9 is a square.

Does this happen again? $x = 7$ and $y = 5$ yield the consecutive pair 49 and 50. The solution yielding the consecutive numbers 8 and 9 differs from the solution yielding the consecutive numbers 49 and 50, since in the first, the square exceeds the "doubled square," while in the second, the doubled square exceeds the square. So we rephrase the question, "when is $x^2 - 2y^2 = \pm 1$?" This happens infinitely often.

Assume that we have a solution, x and y, such that $x^2 - 2y^2 = \pm 1$. Consider the following recursive relation between these known numbers x and y and the new numbers x' and y' given by the equations

$$x' = x + 2y$$
$$y' = x + y$$

Now, observe that $x'^2 - 2y'^2 = (x + 2y)^2 - 2(x + y)^2 = x^2 + 4xy + 4y^2 - 2x^2 - 4xy - 2y^2 = -x^2 + 2y^2 = -(x^2 - 2y^2)$. Since we know that $x^2 - 2y^2 = \pm 1$, it follows that $x'^2 - 2y'^2 = -(\pm 1)$. The point is that this recursive procedure generates infinitely many pairs (x, y) such that $x^2 - 2y^2 = \pm 1$, with each pair larger than the preceding one, since $x' > x$ and $y' > y$.   □

Example: The pair $x = 7$ and $y = 5$ that gave us 49 and 50 yields the new pair $x = 17$ and $y = 12$ that gives us 289 (which is 17^2) and 288 (which is 2×12^2).

The chart below gives the first few pairs generated by the recursive relation.

x	y
1	1
3	2
7	5
17	12
41	29
99	70

An equation of the form

$$x^2 - ky^2 = \pm 1, \tag{2.10}$$

where $k \geq 2$, is called a *Pell* equation. We will see other uses for Pell equations later.

Here is a method for finding infinitely many solutions of (2.10) when the right side is 1.

So now $x^2 - ky^2 = 1$. The first step is to find a pair of positive integers a and b that satisfy this equation, that is, a and b satisfy $a^2 - kb^2 = 1$. Keep in mind that a and b are guesses for x and y. For example,

$x^2 - 2y^2 = 1$ has a solution $x = a = 3$ and $y = b = 2$. Then $3^2 - 2 \cdot 2^2 = 1$. So if n is any positive integer, $(a^2 - kb^2)^n = 1^n = 1$.

Now, suppose that x and y also satisfy $x^2 - ky^2 = 1$. Then for any $n = 1$, 2, 3, 4, ..., we have

$$x^2 - ky^2 = 1 \Rightarrow x^2 - ky^2 = \left(a^2 - kb^2\right)^n \tag{2.11}$$

since $a^2 - kb^2 = 1$.

The next step is to factor both $x^2 - ky^2$ and $a^2 - kb^2$ as the difference of two squares even though k might not be a square. Then (2.11) becomes

$$\left(x - \sqrt{k}y\right)\left(x + \sqrt{k}y\right) = \left(a - \sqrt{k}b\right)^n \left(a + \sqrt{k}b\right)^n$$

Now, one solution of an equation of the form $AB = CD$ can be $A = C$ and $B = D$. Of course, there are many other possibilities. But let's use this approach. This yields the following pair of equations:

$$x + \sqrt{k}y = \left(a + \sqrt{k}b\right)^n$$
$$x - \sqrt{k}y = \left(a - \sqrt{k}b\right)^n$$

To solve for x, add these equations and divide by 2. To solve for y, subtract and divide by $2\sqrt{k}$. We obtain the infinitely many solutions

$$x_n = \frac{\left(a + \sqrt{k}b\right)^n + \left(a - \sqrt{k}b\right)^n}{2}$$
$$y_n = \frac{\left(a + \sqrt{k}b\right)^n - \left(a - \sqrt{k}b\right)^n}{2\sqrt{k}}$$

These equations yield positive integers despite the presence of $\sqrt{k}$. To show this, one applies the binomial theorem. We won't do this here.

My first attempt at producing solutions to the Pell equation followed the procedure in the textbook. There is an issue here regarding fixed place numbers versus floating point numbers. Though the theory says that the procedure will work and produce integers, doing computer arithmetic with limited precision means that it may not work. Java-Script is fine at computing a square root, but it produces a number with a fixed precision (fixed number of digits to the right of the decimal point) and the subsequent computation may end up with numbers such as 3.999999999999. To prevent this problem, I round off the numbers.

My next thought was that it would be possible to guess at the solutions and then check the answers. For a given x, I (my code) solve for the y. I then round off the y and see if it works. Examine each of the programs at the end of the chapter. It may be appropriate to insert bookkeeping code to measure the performances so the two approaches can be compared.

What is the point of all this? Recall that a triangular number has the form $\dfrac{n(n+1)}{2}$. In other words, it is half the product of two consecutive numbers. Consider $(xy)^2$ obtained from any pair (x, y) in the chart. It is clearly a square, but we claim that it is also a triangular number. If so, we have a proof of the following theorem.

Theorem: There are infinitely many numbers that are simultaneously square and triangular.

Proof: Start with a pair of numbers, x and y, satisfying the Pell equation $x^2 - 2y^2 = \pm 1$. Note that there are infinitely many such pairs. Now, the perfect square $(xy)^2$ satisfies

$$(xy)^2 = x^2 y^2 = \frac{x^2 \times 2y^2}{2},$$

which is triangular since $x^2 - 2y^2 = \pm 1$, so the two factors of the numerator are consecutive!   □

The ancient Greeks, by the way, used this chart to approximate $\sqrt{2}$. Dividing both sides of $x^2 - 2y^2 = \pm 1$ by y^2 yields

$$\left(\frac{x}{y}\right)^2 - 2 = \frac{\pm 1}{y^2}$$

As we descend the chart, it is clear that y approaches infinity, implying that the square of its reciprocal goes to 0, that is, $\dfrac{\pm 1}{y^2} \to 0$. Then the left side of the above equation also approaches 0, in which case $\left(\dfrac{x}{y}\right)^2$ approaches 2, implying that the ratio $\dfrac{x}{y}$ approaches $\sqrt{2}$.

Notice that the first five ratios from the chart are 1, 1.5, 1.4, 1.417, and 1.414, rounded to three places. They alternately underestimate and overestimate the actual value. This trend continues forever.

GOLDBACH'S CONJECTURE

Number theory is an old but still vibrant field of mathematics. In the eighteenth century, Goldbach conjectured that every even number greater than 4 is the sum of two odd primes. This has still not been proven! An exercise suggests that you confirm this conjecture by testing the numbers 4 to 1000. (This is not a proof! Something could go wrong for numbers greater than 1000.)

In this programming example, the user is asked to enter a number. The program applies a brute-force approach to generating pairs of primes adding up to the number. The approach is first to generate an array of primes up to a predetermined limit. (See alternative "no-limit" approach described later.) A function, findpairs, uses a for-loop going from 1 to half the chosen number. Let's call the chosen number number. For each value in the iteration, call it f, the numbers f and number-f are each checked as to whether or not they are prime.

Input/output, *aka* user interface operations, are supported in all programming languages but exactly how varies. In this example using JavaScript and HTML, the player's choice of even number is entered using a form and the answers written out in a place in the document, specifically, a div element with id set to "place". You can examine the code to see how this is done and use similar techniques for your own applications. One idiosyncrasy of HTML documents is the default operation of reloading the page. This is why the return false and **return findpairs()** coding is important to prevent the reload from happening and erasing the results.

The example also has some error-checking. If the player enters a number that is not even or a number bigger than the set limit, the alert function is used to display messages.

No-limit: Concerning the preset limit: If you don't want to do this, you can use the code in previous exercises or buried here in the init() function generating the elements in the primes array to check directly if a number is prime. Alternatively, you can use the chosen number to be the limit, generate the primes array up to that value, and then use it.

EXERCISES

An asterisk (∗) indicates that the exercise can be developed into a research project.

1 Find all primes less than 1000. *See if you can do this on your own. Refer to the examples if and when you like. Note: programs can be improved, especially in presentation.*

2 Show that the first 100 primes greater than 3 are either of the form $6k + 1$ or $6k + 5$.

3 Find 100 primes of the form $6k + 5$.

4 Prove the claim that $1^2 + 3^2 + 5^2 + \cdots + (2n+1)^2 = \dfrac{n(4n^2 - 1)}{3}$, by showing that it is true for all $n \le 100$. Now, try to prove it.

5 Prove the claim that $1^3 + 3^3 + 5^3 + \cdots + (2n-1)^3 = n^2(2n^2 - 1)$.

6 Use the chart method to find the sum of the first n fourth powers. Test it for all $n \le 100$.

7 Write a program to add up $f(n)$ for $n = 1, 2, \ldots, N$ where $f(n) = an^3 + bn^2 + cn + d$. The user inputs the integers a, b, c, d, and N. Your program should not employ any of the summation formulas of the chapter. In other words, the program should calculate $f(n)$ for each $n = 1, 2, \ldots, N$ and add the results to a running sum. Then find $\sum_{k=1}^{100} (4k^3 + 6k^2 + 2k + 3)$. Does it matter that the dummy variable is k instead of n?

8 Repeat the previous exercise using the summation laws given by formulas (2.1) and (2.2). How will you handle $cn + d$? Which program is faster for large values of N?

9 ∗13 is the smallest prime such that reversing its digits yields a different prime, namely, 31. Find five more primes with this property.

10 Show that none of the first 100 triangular numbers are the sum of two consecutive squares.

11 Write a program to find the first 50 Fibonacci numbers.

12 Show that the only cubes among the Fibonacci numbers in the previous exercise are 1 and 8.

13 *The Fibonacci numbers 1 and 3 are triangular. Find two more triangular Fibonacci numbers.

14 Verify that for $k \le 100$, u_{3k} is even while the rest are odd. Then prove this for all values of k.

15 Verify that for $k \le 100$, we have $u_1 + u_3 + u_5 + \cdots + u_{2k-1} = u_{2k}$.

16 Verify that for $k \le 100$, we have $u_2 + u_4 + u_6 + \cdots + u_{2k} = u_{2k+1} - 1$.

17 Verify that for $n \le 100$, we have $u_n \le 2^{n-1}$.

18 *Verify that for $n \le 100$, we have $u_n^2 - u_{n+1}u_{n-1} = (-1)^{n-1}$.

19 Find $\sum_{k=1}^{100} \left(4k^3 + 6k^2 + 2k + 3 \right)$, by distributing the sigma and then factoring out the coefficients of each term containing k. Then use a program.

20 Verify that for $n \le 100$, the sum of the first n triangular numbers is $\dfrac{n(n+1)(n+2)}{6}$.

21 *In the eighteenth century, Goldbach conjectured that every even number greater than 4 is the sum of two odd primes. This has still not been proven! Verify the conjecture for the even numbers from 4 to 100. *You can examine the sample program that does the work for a specific number.*

22 Write a program that finds all the ways to write a given even integer $n \ge 6$ as the sum of two odd primes.

23 Verify that every odd number between 9 and 51 is the sum of three odd primes.

24 *The *Ulam sequence* starts with the numbers 1, 2, and 3. Each successive number is the next number that is *uniquely* expressible as the sum of two previous sequence terms. The next number after 1, 2, 3 is 4 since $4 = 1 + 3$ and there is no other way to write 4 as the sum of two of the numbers 1, 2, and 3. The next term after 1, 2, 3, 4 is not 5, since $5 = 1 + 4$ and $5 = 2 + 3$. It is 6, since 6 is uniquely $2 + 4$. Find the first 100 terms of the Ulam sequence.

25 *The *Crunchy sequence* starts with the numbers 1, 2, 3, and 4. Each successive number after 4 is the next number that is *uniquely* expressible as the product of two previous sequence terms. The next number after 1, 2, 3, 4 is 6 since $6 = 2 \times 3$ and there is no other way to write 6 as the product of two of the numbers 1, 2, 3, and 4. The next term after 1, 2, 3, 4, 6 is 8, since $8 = 2 \times 4$. Note that 12 is excluded since $12 = 3 \times 4$ and $12 = 2 \times 6$. The next two terms are 16 and 18. Find the next 50 terms of the Crunchy sequence. Note whether they are prime or composite.

26 *Show that the terms of the Crunchy sequence obtained in the previous exercise have the form $2^n 3^m$.

27 *Prove that the values of x and y in the chart found in this chapter containing solutions to the Pell equation $x^2 - 2y^2 = \pm 1$ (if we extend the chart to produce 30 solutions) have the following properties.

x is always odd.

The parity of y alternates. (y strictly alternates between odd and even values.)

When both of x and y are odd, the integers $\dfrac{x-1}{2}, \dfrac{x+1}{2}$, and y satisfy the Pythagorean theorem, that is, $\left(\dfrac{x-1}{2}\right)^2 + \left(\dfrac{x+1}{2}\right)^2 = y^2$.

x and y have no common factor other than 1. Another way to say this is that x and y are *relatively prime*.

When y is even, let $y = 2k$. Then show that $8k^2 + 1$ is a square.

28 Write a program to find all solutions to the Pell equation $x^2 - 3y^2 = 1$, where x and y are positive integers and $x \le 1000$. *See comment on Exercise 1.*

29 For each $n \le 10{,}000$, determine the number of ways in which n can be written as the sum of two squares. (In some cases, the answer will be zero.) Now, take the average of all your answers.

Basic Program Listing Primes

```
<!DOCTYPE html>
<head>
<meta charset="UTF-8">
<title>Prime number example </title>
<script>

function init(){
for (n=3;n<1000;n++){
  //check if n is a prime
  //first obtain its square root.
  //We make use of a built-in function in JavaScript
```

```
  s=Math.sqrt(n);
  //we have proved n is NOT a prime if we find a factor.
  //we set up what is termed a flag variable. It starts
//out being true.
  noFactorSoFar = true;
  for (f=2;f<=s;f++){
     //is f a factor of n?  We make use of the modulo
//operator, %.
     //It essentially produces the remainder
     r = n % f;
     // r being zero means no remainder, n was
//divisible by f
     if (r==0){
       noFactorSoFar = false;
       break;     // leave the inner f loop because we
//found a factor
     }
     // continue with f loop
  } // ends the f loop
  if (noFactorSoFar){
    //never found a factor
    document.write(n)
    document.write("<br/>");
  }

} // ends the n loop
}
init();

</script>
</head>
<body>
Primes up to 1000
</body>
</html>
```

Basic Program, with Counting of Mod Operation

```
<!DOCTYPE html>
<html>
<head>
<title>Listing primes</title>
<script>
```

```
function init () {
  count = 0;
  np = 0;
  for (n=3;n<1000;n++) {
  //check if n is a prime
  //first obtain its square root.
  //We make use of a built-in function in JavaScript

   s=Math.sqrt(n);
  //we have proved n is NOT a prime if we find a factor.
  //we set up what is termed a flag variable. It starts
//out being true.
   noFactorSoFar = true;
   for (f=2;f<=s;f++) {
      //is f a factor of n?  We make use of the modulo
//operator, %.
      //It essentially produces the remainder
      r = n % f;
      count++;
      // r being zero means no remainder, n was
//divisible by f
      if (r==0) {
        noFactorSoFar = false;
        break;    // leave the inner f loop because we
//found a factor
      }
      // continue with f loop
    } // ends the f loop
    if (noFactorSoFar) {
     //never found a factor
     document.write(n)
     document.write("<br/>");
     np++;
    }
  } // ends the n loop
  document.write("Number of % operations
performed: " + count+"<br/>");
  document.write("Number of primes (starting with
3):"+ np);
}
init();
</script>
```

```
</head>
<body>
<hr/>
Check done using all numbers less than the
square root.
</body>
</html>
```

Improved Program, Just Checking Primes

```
<!DOCTYPE HTML>
<html>
<head>
<title>Faster Listing of Primes </title>
<script>
var primes=[2];   //we treat 2 as a prime
function init(){
     count = 0;
     np=0;
   for (n=3;n<1000;n++){
     //check if n is a prime
     //first obtain its square root.
     //We make use of a built-in function in
//JavaScript

       s=Math.sqrt(n);
     //we have proved n is NOT a prime if we find a
//factor.
     //we set up what is termed a flag variable.
     //It starts out being true.
       noFactorSoFar = true;
     //instead of checking all the numbers, we just
//check the primes
     // the && does a logical AND test BUT
     // it will only calculate the second term
     // if the first is true. This is sometimes
//called "lazy".

       for (fi=0;fi<primes.length && primes[fi]
<=s;fi++){
```

```
      //is f a factor of n?   We make use of the modulo
//operator, %.
      //It essentially produces the remainder
         f=primes[fi];
         r = n % f;
      // r being zero means no remainder, n was
//divisible by f
         count++;
         if (r==0){
           noFactorSoFar = false;
           break;
           // leave the inner f loop because we found
//a factor
         }
      // continue with f loop
      } // ends the f loop
      if (noFactorSoFar){
    //never found a factor
         document.write(n)
         document.write("<br/>");
         primes.push(n);
//add n to our list (array) of primes
         np++; }
    } // ends the n loop


    document.write("Number of % operations
performed: " + count+"<br/>");
    document.write("Number of primes (starting with
3):" + np);
}
init();
</script>
</head>
<body>
<hr/>
Check done by building up list of primes and only
checking for division by primes.


</body>
</html>
```

Basic Solution to a Pell Equation, Using Procedure

```
<!DOCTYPE HTML>
<html>
<head>
<title>Pell solutions k=3 </title>
<script>

function init(){
      //guessed a = 2, b = 1.
   //k is 3
   document.write("Solutions of x² - 3*y² = 1");
   document.write("<hr/>");
   a=2;
   b=1;
   k=3;
   //start with n set to 1. This will produce the x=a
//and y=b values (3,1)
   for(n=1;n<=1000;n++){
      if (produceXandY(a,b,k,n)) break;
//stop when x is over limit

//break out of loop way before n is 1000
   }

}
function produceXandY(a,b,k,n) {
   var x;
   var y;
   var sqrtK = Math.sqrt(k);
   var sqrtKB = sqrtK*b;
   var f1 = a+sqrtKB;
   var f2 = a-sqrtKB;
   var t1n = Math.pow(f1,n);
   var t2n = Math.pow(f2,n);
   x = (t1n + t2n)/2;
   y = (t1n-t2n)/(2*sqrtK);
   //next 2 statements necessary because method
//generates numbers like 2.99999998
   //this is due to limited precision so, for
//examples, values based on sqrt(3) may not cancel
   x = Math.floor(x+.5);
```

```
  y = Math.floor(y+.5);
  if (x>1000) {return true;}
  else {
  document.write("x = "+ String(x)+ ", y = "+String
(y)+ "<br/>");
  return false;
   }
}
init();
</script>
</head>
<body>
<hr/>
Check done based on "guess" of x=a=3 and y=b=1 and
then using procedure to iterate

</body>
</html>
```

Solving a Pell Equation by Direct Calculation and Checking if Answer Is an Integer

```
<!DOCTYPE HTML>
<html>
<head>
<title>Pell solutions k=3 crude</title>
<script>
function init(){

  //k is 3
  document.write("Solutions of x² - 3*y² = 1");
  document.write("<hr/>");

  k=3;

  //start with n set to 1. This will produce the x=a and
  //y=b values (3,1)
  for(x=1;x<1000;x++){
  h = (x*x-1)/k;
  yy = Math.sqrt(h);
  y  = Math.floor(yy+.5);
```

```
  //idea is that the rounded off integer value
//won't work
  if ((x*x-3*y*y)==1) {
  document.write("x = "+ String(x)+ ", y = "+String
(y)+ "<br/>");
   }
   }
}
init();
</script>
</head>
<body>
<hr/>
Crude procedure checking for each value of x, the
computed value of y, taking the positive square root
</body>
</html>
```

Interactive Program to Generate Pairs of Primes Adding up to Even Number

```
<!DOCTYPE HTML>
<html>
<meta charset="UTF-8">
<head>
<title>Numbers as sum of 2 primes </title>
<script>
var primes=[2];   //we treat 2 as a prime
var limit = 1000;
function init(){
  var f;
  var r;
  var noFactorSoFar;
  var s;
  for (var n=3;n<limit;n++){
  //check if n is a prime for each n
  //first obtain its square root. We make use of a
//built-in function in JavaScript

    s=Math.sqrt(n);
  //we have proved n is NOT a prime if we find a factor.
```

```
    //we set up what is termed a flag variable. It starts
//out being true.
     noFactorSoFar = true;
   //instead of checking all the numbers, we just
//check the primes
   // the && does a logical AND test BUT it will only
//calculate the second term
   // if the first is true. This is sometimes
//called "lazy".
     for (var fi=0;fi<primes.length && primes[fi]
<=s;fi++){
      //is f a factor of n?   We make use of the modulo
//operator, %.
      //It essentially produces the remainder
        f=primes[fi];
        r = n % f;
      // r being zero means no remainder, n was
//divisible by f
        if (r==0){
          noFactorSoFar = false;
          break;
// leave the inner f loop because we found a factor
        }
     // continue with fi loop
    } // ends the fi loop
     if (noFactorSoFar){
    //never found a factor
        primes.push(n);   of primes

    }} // ends the n loop

}
function findpairs() {
//add n to our list (array) of primes
  answer = "";
//will use to construct the answer
  count = 0;   var h;
  var number = parseInt(document.f.num.value);
//extract what the user entered
  if ((number%2)>0) {
    alert("Your number must be even. Try again");
    return false;}
```

```
  if (number > limit) {
    alert("Your number is greater than "+limit+".
Please try a smaller number.");
    return false;
  }
  h = number / 2;
  for (f=1;f<=h;f++) {
    //need to check if f and (number-f) both primes.
//Use lazy evaluation
    if (isPrime(f) && isPrime(number-f)) {
      answer += String(f)+ " and "+ String(number-f)
+" <br/>";
      count++;
    }
  }

  if (answer.length>0) {   //were any pairs found?
     answer +=String(count)+" pairs of primes
adding up to "+String(number);

  }
  else {   //this may never happen???
     answer = document.createTextNode("No pairs of
primes adding up to "+String(number));
  }
  document.getElementById("place").innerHTML =
answer;
  return false;
// required to prevent the browser from re-loading
original document
}

function isPrime(c) {
  var i;
  i = primes.indexOf(c);
  return (i>=0);
//i greater or equal to zero means that c is in the
primes array

}
</script>
</head>
```

```
<body onload="init();">
Find pair of prime numbers adding up to even number
<br/>
<form name="f" onsubmit="return findpairs();">
  <input type="text" name="num" value="  "/>
  <input type="submit" value="Submit an even
number "/>
</form>
<p id="place">
  The answer will go here.
</p>
</body>
</html>
```

3

PASCAL'S TRIANGLE

In this chapter, we discuss Pascal's triangle and explain the relevance of its entries to number theory.

FACTORIALS

As you are probably aware, $n!$, or *n factorial*, is the product of the first n positive integers, that is,

$$n! = n \times (n-1) \times (n-2) \times \cdots \times 3 \times 2 \times 1$$

We define $0!$ to be 1. There are $n!$ permutations of n distinct objects.

There are two ways (at least) of programming factorial. One uses the definition: factorial of n is the product of the first n integers. This is called the iterative way. It is not too unsurprising that this is coded using a for-loop, going from 1 to n. The exact coding is

```
var ans = 1;
    for (var j=1;j<(n+1);j++){
```

Elementary Number Theory with Programming, First Edition. Marty Lewinter and Jeanine Meyer.
© 2016 John Wiley & Sons, Inc. Published 2016 by John Wiley & Sons, Inc.

```
    ans *= j;
}
```

The second way uses recursion. The informal definition of a recursion function is one that calls itself. (A critical concept in theoretical computer science is that the recursive functions are the set of functions obtained when starting with a small set of functions and allowing certain function making procedures including composition and finding the inverse. One of the function making procedures is termed primitive recursion.) Using recursion, factorial is

If n ≤ 1, factorial(n) is 1

Otherwise, factorial(n) is n∗factorial(n − 1)

(This definition is written in what is termed pseudocode, a combination of coding and English.)

This definition can be used as the basis of a program. This may be surprising and the technique can be misused. The programmer has to be sure that the recursion will end; that is, that it isn't equivalent to circular reasoning. In this particular case, observe that the value passed as the parameter to the factorial function is going down and so will eventually reach 1. The relevant code in my program is

```
function worker(n) {
    if (n<=1) { return 1; }
    else {return n*worker(n-1);}
}
```

So now let's ask the question, which way is better? That is, which is faster? The answer was slightly different than I anticipated. Before giving away the punch line, let me say why I thought that iteration would be better. Recursion is a wonderful technique. However, certain operations are performed when we use recursion. There is something called the program stack, and information about each function call is *pushed* onto the stack and used until the function completes and then it is *popped*, removed from the stack. My thinking was that the bookkeeping for a for-loop is less computationally heavy than the bookkeeping for function calls.

I added code to keep track of time, using Date, and getTime, as was done in earlier examples. By the way, I needed to extract the functionality that displayed the answers from the inner program.

That is why you see a function called `worker` in my example. You can examine both programs at the end of the chapter.

It was the case that at some point (you figure out when), there was an error message of *too much recursion*! However, before that point, for both the iterative approach and the recursion approach, something else broke down! The numbers grew too big. Recall scientific notation. The number 123 can be expressed as 1.23 times 10^2. It also can be expressed as 12.3 times 10^1 or 123 times 10^0.

This is called floating point in programming. All numbers in JavaScript are kept using what is termed *double precision*, using 64 bits, each bit being a 1 or a 0. For the 64 bits, using the numbering system 0 to 63, bits 0 to 51 hold a number, bits 52 to 62 hold an exponent, and bit 63 is used for a sign (0 for positive and 1 for negative). Now, returning to example above with 123 (which conveniently uses decimal, i.e., base 10), let's say our system only allowed three decimal spaces for the number. Our system could not precisely express 1234! Let's say our system only allowed the exponent to be 0 to 9. Then the system could not express small numbers such as .12 or very big ones. JavaScript is better than what I just described, but if and when a number exceeds its capacity in one of these ways, JavaScript says the number is infinity! Try it. As it turns out, though iterative factorial does not produce the *too much recursion* error, the fact that the number is outside the capacity of double precision occurs the same time for both methods.

If you have an application in which double precision is not good enough, then you can try and find a library for arbitrary or unlimited precision. Such libraries exist and we may use one for a later topic.

THE COMBINATORIAL NUMBERS *n* CHOOSE *k*

Let us consider the following problem. Given n distinct objects, in how many ways can we select k of them where order does not matter? If I have three extra tickets to a baseball game and I have 10 friends, in how many ways can I select the lucky trio of friends who will receive free tickets? Here, order does not matter. The choice Newton,

Archimedes and Euclid is the same as the choice Euclid, Newton and Archimedes.

We begin with the first k factors of $n!$, namely, $n \times (n-1) \times (n-2) \times \cdots \times (n-k+1)$. This is because there are n ways to pick the first object, $n-1$ ways to pick the second object, and so on till the last choice (the kth object), which can be made in $(n-k+1)$ ways.

The expression $n \times (n-1) \times \cdots \times (n-k+1)$ is too big to be the answer to our problem. To see this, consider the number of combinations of three objects from among six distinct objects. Let the six objects be the letters A through F. The three-letter choice ACD, for example, will be counted six times instead of once. It will appear in the list of permutations as ACD, ADC, CDA, CAD, DCA, and DAC. But we only want it (and all the other combinations such as ACE, etc.) listed *once* in the list of combinations. So we divide by 6—the number of permutations of three objects. This suggests the following procedure for the number of combinations of k objects chosen from among n distinct objects. Divide the product $n \times (n-1) \times (n-2) \times \cdots \times (n-k+1)$ by $k!$. In symbols,

$$\binom{n}{k} = \frac{n \times (n-1) \times (n-2) \times \cdots \times (n-k+1)}{k!} \tag{3.1}$$

The symbol on the left of Equation 3.1 is pronounced "n choose k." If we multiply the numerator and denominator by $(n-k)!$, we obtain the alternative formula

$$\binom{n}{k} = \frac{n \times (n-1) \times (n-2) \times \cdots \times (n-k+1) \times (n-k)!}{k!} = \frac{n!}{k!(n-k)!} \tag{3.2a}$$

The numerator of the middle expression in (3.2a) is seen to be $n!$ after we expand the $(n-k)!$, thereby realizing it supplies all the factors of $n!$ that were missing in the numerator of (3.1).

Let's rewrite (3.2a) without the middle expression:

$$\binom{n}{k} = \frac{n!}{k!(n-k)!} \tag{3.2b}$$

The numbers k and $n-k$ in the denominator of (3.2b) are called *complements* because they add up to n. If we replace each k by its complement, $n-k$, we obtain

$$\binom{n}{n-k} = \frac{n!}{(n-k)![n-(n-k)]!} = \frac{n!}{(n-k)!k!} = \binom{n}{k}$$

or, more simply,

$$\binom{n}{n-k} = \binom{n}{k} \qquad\qquad (3.3)$$

After all, choosing a meal of k dishes from a menu of n is equivalent to indicating the $n - k$ dishes not chosen. Thus, to evaluate $\binom{100}{97}$, it is easier to evaluate $\binom{100}{3} = \frac{100 \times 99 \times 98}{3 \times 2 \times 1}$.

Observe that "n choose n" yields two answers, which may be compared. On the one hand, it is, using formula (3.2b), $\frac{n!}{0!n!}$, while on the other hand, common sense tells us that the answer must be 1. (There is only one way to choose all n objects.) It follows that 0! must be 1. Moreover, it follows that $\binom{n}{0} = 1$.

Another useful observation is that $\binom{n}{1} = n$.. This could, of course, be verified without the help of Equation 3.1 by using common sense. There are n ways to choose one item among n distinct items.

PASCAL'S TRIANGLE

There is an ancient array of numbers called *Pascal's triangle*, shown below up to the seventh row. (The name of the row is given by the second entry, printed in bold.)

```
                    1
                 1     1
              1     2     1
           1     3     3     1
        1     4     6     4     1
     1     5    10    10     5     1
  1     6    15    20    15     6     1
1     7    21    35    35    21     7     1
```

Pascal's Triangle

Each entry is the sum of the two entries immediately above it. Thus, the first 35 in the seventh row is the sum of the 15 and 20 in the sixth row.

You should have no trouble obtaining the entries of the eighth row using this method of generation.

> When first confronting the exercise, I admit that I thought of using the factorial and the *n choose k* formulas as described next in this chapter. However, I quickly decided that the basic definition could be a sound alternative. That is, each row of the triangle is constructed from the last. If I could keep track of positions, it would be a good way.
>
> How can I (my code) produce the next row? The first step is to take a row and add in what I view as the ghost elements, a zero at the start and a zero at the end. If the row is
>
> 1 3 3 1
>
> then my code constructs an array of elements: 0, 1, 3, 3, 1, 0.
>
> The next row will be
>
> 0 + 1, 1 + 3, 3 + 3, 3 + 1, 1 + 0.
>
> This *next* row is one longer than the original row. The challenge is to produce the extended row. As it turns out, JavaScript has two array methods that do exactly what is called for: unshift adds a new element at the start and push adds a new element at the end. These methods (*method* is the technical term for a function associated with a given *class* of objects, in this case, with arrays) actually change the array and return the length of the new array. I use this in the code:
>
> ```
> count = last.unshift(0);
> ```
>
> Another challenge of the coding is that I use a for-loop inside a while loop. A variable called limit is used to keep count of the number of rows remaining to be done. You can examine the code at the end of the chapter.
>
> One last point, I did have a worry about formatting the triangle nicely, but it came almost for free. I am using JavaScript with html and html has <center> and </center> tags. My code creates a long string of characters. The code turns the numbers into character strings using String and puts a space between numbers using the   entity. It stands for nonbreaking space. A line break is indicated by the tag <br/>. Centering centers each line.

BINOMIAL COEFFICIENTS

Now, it is amazing that the entries of the nth row of Pascal's triangle consist of the $n + 1$ combinatorial numbers:

$$\binom{n}{0}, \binom{n}{1}, \binom{n}{2}, \dots, \binom{n}{n-2}, \binom{n}{n-1}, \binom{n}{n}$$

These combinatorial numbers are the coefficients in the expansion of $(a + b)^n$ and are often called *binomial coefficients*. Moreover, the third entry of each row, that is, the numbers of the form $\binom{n}{2} = \frac{n(n-1)}{2}$, are the sums of the positive integers from 1 to $n - 1$, as can be seen by using equation 1.2 in Chapter 1, after replacing n by $n - 1$. These numbers are triangular.

The principle used to generate the $(n + 1)$st row from the nth row can be written as

$$\binom{n+1}{k+1} = \binom{n}{k} + \binom{n}{k+1} \tag{3.4}$$

Theorem: The sum of the entries of the nth row of Pascal's triangle is 2^n.

Theorem: $\binom{n}{0}^2 + \binom{n}{1}^2 + \binom{n}{2}^2 + \cdots + \binom{n}{n-2}^2 + \binom{n}{n-1}^2 + \binom{n}{n}^2 = \binom{2n}{n}$.

EXERCISES

An asterisk (*) indicates that the exercise can be developed into a research project.

1 Write a program to find $n!$ for a given nonnegative integer, n. Then use this program to tabulate $n!$ for all $n \le 20$. *You can use either of the factorial examples to help you do this.*

2   Using the previous exercise, write a program to find $\binom{n}{k}$ for a given pair of integers k and n such that $0 \leq k \leq n/2$, with $n \neq 0$, using formula (3.2). (Why can we assume that $k \leq n/2$?)

3   Now, write a program to find $\binom{n}{k}$ using formula (3.1). Which program is faster? Why?

4   Write a program that prints out the first k rows of Pascal's triangle for a given k. *See the example at the end of the chapter. Perhaps you can improve the formatting.*

5   $10! = 6!7!$. Are there more relations of the form $c! = a!b!$ where $2 \leq a \leq b < c \leq 30$?

6   *The 14th oblong number, 210, can be written as $5 \times 6 \times 7$. Are there other oblong numbers less than 1000 that can be written as products of three consecutive numbers?

7   *Find all triangular numbers less than 1000 that can be written as products of three or more consecutive integers.

8   *Show that $\binom{n}{1} + 2\binom{n}{2} + 3\binom{n}{3} + \cdots + n\binom{n}{n} = n2^{n-1}$ for $n \leq 100$.

9   The *Catalan numbers* are defined by $c_n = \dfrac{1}{n+1}\binom{2n}{n}$ and are useful in combinatorics, number theory, and computer science. They satisfy the recurrence relation $c_n = \dfrac{2(2n-1)}{n+1}c_{n-1}$. Find the first 100 Catalan numbers (i) using this recurrence and (ii) using the definition. Which is faster?

10 For $n \leq 1000$, show that the entries of the nth row of Pascal's triangle, excluding the initial and final 1's, are all even if and only if n is a power of 2. The entries of the fourth row, for example, are 1, 4, 6, 4, 1, which, except for the initial and final 1's, are even.

11 For $n \leq 100$, show that $(n)1 + (n-1)2 + (n-2)3 + \cdots + 3(n-2) + 2(n-1) + 1(n) = \binom{n+2}{3}$.

12 For $n \le 100$, show that $1 \times 2 \times 3 + 2 \times 3 \times 4 + \cdots + n(n+1)(n+2) = \dfrac{n(n+1)(n+2)(n+3)}{4} = 6\dbinom{n+3}{4}$.

13 *Verify that the following infinite array of equations involving the triangular numbers is correct for $n \le 100$.

$$t_1 + t_2 + t_3 = t_4$$

$$t_5 + t_6 + t_7 + t_8 = t_9 + t_{10}$$

$$t_{11} + t_{12} + t_{13} + t_{14} + t_{15} = t_{16} + t_{17} + t_{18}$$

$$t_{19} + t_{20} + t_{21} + t_{22} + t_{23} + t_{24} = t_{25} + t_{26} + t_{27} + t_{28}$$

$$\vdots$$

14 Find 50 primes of the form $5n + 1$.

15 Given the polynomial $f(n) = n^2 + n + 41$, verify that $f(n)$ is prime for $n = 0, 1, 2, \ldots, 39$.

16 Show that $n!$ is not a square when $1 < n < 50$.

17 *Consider the equation $n!m! = s!t!$ where $m > n$ and $t > s$. Write a program that accepts the values of n and m from the user and finds a pair s and t or determines that none exists. Verify that $4!15! = 7!13!$ yields an example of this equation.

Iterative Factorial

```
<!DOCTYPE HTML>
<html>
<meta charset="UTF-8">
<head>
<title>Factorial Iterative</title>
<script>
function fac(ns) {
  var n = parseInt(ns);
  var start = new Date();
  var startmm = start.getTime();
  var ans = 1;
  for (var j=1;j<(n+1);j++) {
    ans *= j;
  }
  var now = new Date();
  nowmm = now.getTime();
```

```
  var elapsed = nowmm - startmm;
  alert("ans is "+ans+" performed in "+elapsed
  +" msec.");
  return ans;
}

</script>
</head>
<body>
Choose limit <br/>
<form name="f" onsubmit="return fac(this.
num.value);">
  <input type="text" name="num" value="  "/>
  <input type="submit" value="Submit a number "/>
</form>
<p id="place">
  The answer will go here.
</p>
</body>
</html>
```

Recursive Factorial

```
<!DOCTYPE HTML>
<html>
<meta charset="UTF-8">
<head>
<title>Factorial Recursive</title>
<script>
function fac(ns){
  var n = parseInt(ns);
  var start = new Date();
  var startmm = start.getTime();
  var ans = worker(n);
  var now = new Date();
  nowmm = now.getTime();
  var elapsed = nowmm - startmm;
  alert("ans is "+ans+" performed in "+elapsed
  +" msec.");
  return ans;
}
function worker(n){
```

```
  if (n<=1) { return 1;}
  else {return n*worker(n-1);}
}

</script>
</head>
<body>
Choose limit <br/>
<form name="f" onsubmit="return fac(this.
num.value);">
  <input type="text" name="num" value="  "/>
  <input type="submit" value="Submit a number "/>
</form>
<p id="place">
  The answer will go here.
</p>
</body>
</html>
```

Pascal's Triangle

```
<title>Pascal's Triangle</title>
<script>

function triangle(ns) {
  var limit = parseInt(ns);
  var output = "<center>";
  output +="1<br/>";
  var last = [1];
  var lastext;
  var next;
  var val;
  limit--;
  while (limit>=0) {
    //unshift and push each change the array itself.
    //Each returns new length
      count = last.unshift(0);
      last.push(0);
      next = [];

      for (var i=0;i<count;i++) {
          val = last[i]+last[i+1];
```

```
            next.push(val);
            output+=String(val)+" ";
    }
    output+="<br/>";

    last = next;
    limit--;
  }

  output+="</center>";
  document.getElementById("place").innerHTML =
output;
  return false;
}
</script>
</head>
<body>
Choose limit <br/>
<form name="f" onsubmit="return triangle(this.
num.value);">
  <input type="text" name="num" value="  "/>
  <input type="submit" value="Submit a number "/>
</form>
<p id="place">
  The answer will go here.
</p>
</body>
</html>
```

4

DIVISORS AND PRIME DECOMPOSITION

We present an ancient theorem on the representation of positive integers as products of prime numbers. The consequences of this theorem are important for a host of reasons.

DIVISORS

A factor of a number is called a *divisor*. Thus, 10 has the divisors 1, 2, 5, and 10. The divisors of *n* other than *n* are called *proper* divisors. The proper divisors of 10 are, therefore, 1, 2, and 5, while its divisors are 1, 2, 5, and 10. A divisor of *n* is said to *divide n*. Thus, 5 divides 10, while 7 does not divide 10. The symbol for "divides" is a vertical line. When *a* divides *b*, we write *a* | *b*. Whenever *a* divides *b*, observe that *b* is a *multiple* of *a*, that is, $b = ka$, for some integer *k*. Thus, $5 \mid 10$, while $10 = 2 \times 5$. In fact, this is an "if and only if" statement:

a divides *b* if and only if *b* is a multiple of *a*.

In symbols, we have, using the two-way implication arrow,

$$a|b \Leftrightarrow b = ka \qquad (4.1)$$

Elementary Number Theory with Programming, First Edition. Marty Lewinter and Jeanine Meyer.
© 2016 John Wiley & Sons, Inc. Published 2016 by John Wiley & Sons, Inc.

Here is a nice way to think about (4.1). Even though 20 divides 100, one may write $100 = 4 \times 25$, preventing us from seeing the 20. On the other hand, we can write $100 = 5 \times 20$, thereby showing us the divisor (or factor) 20.

Example: Let $N = 17 \times 31 \times 101$. Show that $31 \mid N$. While this seems obvious enough, let's use (4.1). Let $k = 17 \times 101$. Then $N = 31k$ in which case $31 \mid N$. I had a student who did this problem by multiplying 17, 31, and 101 to find N and then dividing N by 31 yielding no remainder.

Example: Let p be prime. If $a \mid p$, what can be said of a? Since p is prime, $a = 1$ or p.

Example: Find a number with exactly three divisors. Then find infinitely many such numbers. To begin, the divisors of 4 are 1, 2, and 4. Other examples include 9 (1, 3, and 9) and 25 (1, 5, and 25). These examples are squares. But 16 is a square whose divisors are 1, 2, 4, 8, and 16. Thus, it doesn't suffice to be a square. What about squares of primes? If p is prime, the divisors of p^2 are 1, p, and p^2. Aha! We have our so-called infinite class of solutions. (In fact, we have all solutions. Think about it.) How many divisors does p^4 have?

Here are a few laws of divisibility. As is our practice throughout the text, all letters represent integers (whole numbers) unless otherwise stated:

1. $a \mid b$ and $b \mid c \Rightarrow a \mid c$.
2. $a \mid b$ and $b \mid a \Leftrightarrow a = \pm b$.
3. $a \mid b$ and $c \mid d \Rightarrow ac \mid bd$.
4. $a \mid b \Rightarrow ac \mid bc$ (where $c \neq 0$).
5. $a \mid r$ and $a \mid s \Rightarrow a \mid (mr + ns)$.
6. $a \mid r$ and a does not divide $s \Rightarrow a$ does not divide $r + s$.
7. If a and b are positive integers such that $a \mid b$, then $a \leq b$.

We shall prove Law 1 and leave the rest for the reader. Using (4.1), we have $b = ka$ and $c = jb$. Then we obtain the following chain of equations: $c = jb = jka = (jk)a = ha$, where $h = jk$. Equating the first and last parts of this super equation, we get $c = ha$. Then by (4.1), $a \mid c$. $\qquad\square$

The expression $mr + ns$ in Law 5 is called a *linear combination of r and s*, since it is the sum of *multiples* of r and s. So 530 is a linear combination of 100 and 10, namely, $5 \times 100 + 3 \times 10$. This concept can be extended to more than two numbers. $mr + ns + pt$ is a linear combination of r, s, and t. So 534 is a linear combination of 100, 10, and 1, namely, $5 \times 100 + 3 \times 10 + 4 \times 1$. By contrast, expressions such as $6r^2 + 10s^3$, $2rs$, $\sqrt{r} + \dfrac{1}{s}$, are nonlinear.

Example: Show that if $a \mid b$ and $c \mid d$, it does not follow that $(a + c) \mid (b + d)$. Let $a = 2$, $b = 8$, $c = 3$, and $d = 9$. Then $2 \mid 8$ and $3 \mid 9$, while 5 does not divide 17.

Let's prove by induction that $8 \mid (25^n + 7)$. The base case, $n = 1$, is obvious since $8 \mid 32$. Now, we assume that $8 \mid (25^k + 7)$ and we must show that $8 \mid (25^{k+1} + 7)$. We have, using (4.1), that $25^k + 7 = 8j$. Then $25^k = 8j - 7$. Multiplying by 25 yields $25^{k+1} = 25(8j - 7) = 200j - 175$. Adding 7 to both sides yields $25^{k+1} + 7 = 200j - 168 = 8(25j - 21)$. After eliminating the middle member of this equation, we have $25^{k+1} + 7 = 8(25j - 21)$, showing that 8 is indeed a divisor of $25^{k+1} + 7$, and the clever proof is over.   □

Imagine how difficult it would be to compute $25^{837} + 7$ in order to show that it is a multiple of 8.

GREATEST COMMON DIVISOR

Given a and b, we call d the *greatest common divisor* of a and b, denoted $gcd(a,b)$, if:

1. $d \mid a$ and $d \mid b$.
2. d is the greatest number that divides both a and b.

Example: $gcd(4,6) = 2$, $gcd(15, 45) = 15$, and $gcd(18, 25) = 1$. Note that while $3 \mid 15$ and $3 \mid 45$, $gcd(15, 45) \neq 3$ because $5 > 3$ and 5 is also a common divisor of 15 and 45.

If $gcd(a,b) = 1$, we say that a and b are *relatively prime*, that is, they have no common divisor other than 1. Note from the preceding example that

$$gcd(a,b) \leq min\{a,b\}$$

where $min\{a,b\}$ denotes the smaller of a and b. Furthermore, we will have equality if and only if $a \mid b$ (as in our second example, $gcd(15,45) = 15$).

It is important to note that $gcd(a,b) = d$ if and only if $a = rd$ and $b = sd$, where r and s are relatively prime. To illustrate this, observe that $15 = 3 \times 5$ and $35 = 7 \times 5$. Since 3 and 7 are relatively prime, we see that $gcd(15,35) = 5$. Note, also, that when we reduce a fraction to lowest terms, we cancel the gcd of the numerator and denominator, leaving a new pair of integers with a gcd of 1. In the notation of the first sentence of this paragraph, we have $\dfrac{a}{b} = \dfrac{rd}{sd} = \dfrac{r}{s}$. The last fraction is reduced to lowest terms since, by assumption, $gcd(r,s) = 1$. So $gcd(a,b)$ is the greatest number that we can cancel when we reduce $\dfrac{a}{b}$. Furthermore, $gcd(a,b) = d$ is the greatest number that we can factor out of the expression $a + b$, that is, $a + b = d(r + s)$. For example, $100 + 175 = 25(4 + 7)$. Since 4 and 7 are relatively prime, we can't factor further.

Before we present an algorithm for finding $gcd(a,b)$, here is the way we write the division of b by the positive number a in a way that avoids fractions. (We wish to avoid statements such as $\dfrac{7}{2} = 7 \div 2 = 3\frac{1}{2}$.)

Division: Given integers a and b, where a is positive, there exists a pair of numbers, q (for *quotient*) and r (for *remainder*), where $0 \le r < a$, such that $b = qa + r$.

If $a = 4$ and $b = 35$, for example, then $q = 8$ and $r = 3$. We have $35 = 8 \times 4 + 3$. In words, this says that 35 divided by 8 is 4 and the remainder is 3. Note that $3 < 4$. This is why $r < a$. Since we are dividing by a, the remainder must be less than a, or we would get a larger quotient. If one claims that $20 \div 6 = 2$ with a remainder of 8, we would protest since 6 goes into 8 one time (with a remainder of 2), implying that $20 \div 6 = 3$ with a remainder of 2. Note, furthermore, that $r = 0$ if and only if $a \mid b$ or, equivalently, $r = 0$ if and only b is a multiple of a. In this case, $b = qa$, or equivalently, $\dfrac{b}{a} = q$.

Example: Given $a = 7$ and $b = 66$, find q and r, where $0 \le r < 7$, such that $66 = 7q + r$. Since $66 \div 7 = 9$ with a remainder of 3, we have $q = 9$ and $r = 3$. Then $66 = 7 \times 9 + 3$.

Note that if d is a common divisor of a and b, then given any number n, it follows that d must also divide $b - na$, by Law 5 above. For instance, $d = 5$ is a common divisor of $a = 15$ and $b = 100$. Then 5 is also a common divisor of 15 and 85, which is $100 - 15$ or $b - 1a$. It is also a common divisor of 15 and 70, which is $100 - 2 \times 15$ or $b - 2a$, etc.

Moreover, if $d \mid a$ and $d \mid (b - na)$, then $d \mid b$, since $b = na + (b - na)$. In other words, d is a common divisor of a and b if and only if d is a common divisor of a and $b - na$ for any n. Stated formally,

$d = gcd(a,b)$ if and only if $d = gcd(a, b - na)$, for any integer n

This last calculation will be expedited by making $b - na$ as small as possible using the division algorithm. This is done by letting $n = q$, where q is the quotient when b is divided by a. Then $b - qa$ will be the remainder r in the equation $b = qa + r$. It then follows that $b - qa < a$. The point of all this is that $gcd(a,b) = gcd(a, b - qa)$.

Example: To find $gcd(20,110)$, observe that $110 = 5 \times 20 + 10$. Then $gcd(20,110) = gcd(10,20)$, which is indeed easier to compute. Of course, now we can reapply the algorithm after observing that $20 = 2 \times 10 + 0$, implying that $gcd(10,20) = min\{10,20\} = 10$.

Example: To find $gcd(30,84)$, we observe that $84 = 2 \times 30 + 24$. Then it suffices to compute $gcd(24,30)$. Observe, first, that $30 = 1 \times 24 + 6$, in which case it suffices to compute $gcd(6,24)$. Observing that $24 = 4 \times 6 + 0$, we find that the gcd is 6. The work can be organized as follows:

$$84 = 2 \times 30 + 24$$
$$30 = 1 \times 24 + 6$$
$$24 = 4 \times 6 + 0$$

The remainder in the second to the last equation yields the gcd, 6, of 30 and 84. ("Second to the last" can be shortened to "penultimate." Not bad for a math writer, eh?)

Example: Find $d = gcd(30,292)$. We obtain the following equations:

$$292 = 9 \times 30 + 22$$
$$30 = 1 \times 22 + 8$$
$$22 = 2 \times 8 + 6$$
$$8 = 1 \times 6 + 2$$
$$6 = 3 \times 2 + 0$$

We conclude from the next to the last equation, $8 = 1 \times 6 + 2$, that $d = 2$.

The final "division" equation will always have a zero remainder, since the remainders form a strictly decreasing sequence of nonnegative numbers. The remainder in the second to the last equation in the example above yields the *gcd*, 2, of 30 and 292. In general, to find $d = gcd$ (a,b), we will obtain a sequence of equations of the form

$$b = q_1 a + r_1 \quad \text{where } r_1 < a$$
$$a = q_2 r_1 + r_2 \quad \text{where } r_2 < r_1$$
$$r_1 = q_3 r_2 + r_3 \quad \text{where } r_3 < r_2$$
$$\vdots$$
$$r_{k-3} = q_{k-1} r_{k-2} + r_{k-1} \quad \text{where } r_{k-1} < r_{k-2}$$
$$r_{k-2} = q_k r_{k-1} + r_k \quad \text{where } r_k < r_{k-1}$$
$$r_{k-1} = q_{k+1} r_k + 0 \quad \text{where } 0 < r_k$$

and $r_k = d = gcd(a,b)$. Note that $r_1 > r_2 > r_3 > \ldots > r_k = d$.

The greatest common divisor algorithm suggests using recursion for the program. That is, the $gcd(a,b)$ is the same as $gcd(a, b-qa)$, where q is the quotient of b divided by a. We use the modulo operator of JavaScript, %, to compute $b-qa$:

```
r = b % a;
```

This satisfies the requirement for successful use of recursion: the parameter values are decreasing.

There is some preliminary work we need to do. The program is set up to be interactive with the user typing in the two numbers. The input must be converted to be integers. My code also makes sure that the "a" value is the smaller of the two inputs.

I decided that my program would display messages indicating the process of the algorithm. Here is a screenshot from the program.

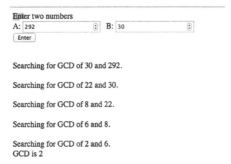

In programs such as these, it can be tricky to start. The input form calls on one program, `find`, that sets up the reference variable for the place for the messages and extracts the input values, converting each to integers. The `gcd` program does the work, including invoking itself when appropriate. The complete program is at the end of the chapter.

The Euclidean algorithm yields an interesting fact, which we will find useful. The second to the last equation can be written (recalling that $d = r_k$)

$$d = r_{k-2} - q_k r_{k-1} \tag{4.2}$$

From the equation before that one, we get $r_{k-1} = r_{k-3} - q_{k-1} r_{k-2}$, which allows us to rewrite (4.2) as $d = r_{k-2} - q_k(r_{k-3} - q_{k-1} r_{k-2}) = -q_k r_{k-3} + r_{k-2} + q_k q_{k-1} r_{k-2}$ or

$$d = -q_k r_{k-3} + (1 + q_k q_{k-1}) r_{k-2} \tag{4.3}$$

Using the equation, $r_{k-4} = q_{k-2} r_{k-3} + r_{k-2}$, we can replace r_{k-2} in (4.3) by $r_{k-4} - q_{k-2} r_{k-3}$, obtaining

$$d = -q_k r_{k-3} + (1 + q_k q_{k-1})(r_{k-4} - q_{k-2} r_{k-3}) = A r_{k-4} + B r_{k-3}, \tag{4.4}$$

where A and B are expressions involving q_{k-2}, q_{k-1}, and q_k.

What is the point? Moving up the ladder using the above procedure, we will finally obtain $d = ma + nb$, where m and n are integers! In other words, $gcd(a,b)$ is a linear combination of a and b.

In fact, d is the smallest *positive* number expressible as a linear combination of a and b. Here's why. Assume to the contrary that $0 < c < d$ and $c = ra + sb$. Now, $d \mid (ra + sb)$, so $d \mid c$, which is absurd.

Note in particular that if $d = gcd(a,b) = 1$, that is, if a and b are relatively prime, then there are integers m and n such that $ma + nb = 1$.

Example: To express $gcd(30,84) = 6$ as a combination of 30 and 84, the above procedure suggests that we start with the Euclidean algorithm, as we did earlier:

$$84 = 2 \times 30 + 24$$
$$30 = 1 \times 24 + 6$$
$$24 = 4 \times 6 + 0$$

The second equation implies that $6 = 30 - 24$. By the first equation, $24 = 84 - 2 \times 30$. Substituting the right side of this equation for 24 in $6 = 30 - 24$ yields $6 = 30 - (84 - 2 \times 30) = 3 \times 30 - 1 \times 84$.

Example: Prove that $gcd(9n + 4, 2n + 1) = 1$. In other words, show that $9n + 4$ and $2n + 1$ are relatively prime for any positive n. Using the Euclidean algorithm, we have

$$9n + 4 = 4 \times (2n + 1) + n$$
$$2n + 1 = 2 \times n + 1$$
$$n = n \times 1 + 0$$

proving the bold assertion!

Example: Prove that $gcd(n^2 - n + 1, n + 1) = 1$ or 3. We have $n^2 - n + 1 = n^2 - n - 2 + 3 = (n - 2)(n + 1) + 3$, in which case $gcd(n^2 - n + 1, n + 1) = gcd(n + 1, 3)$, which is either 1 or 3. Note that if $3 \mid (n + 1)$, it follows that $3 \mid (n - 2)(n + 1) + 3$. Then $gcd(n^2 - n + 1, n + 1) = 3$.

Example: Let p be prime and let $d = gcd(p, a)$. Show that $d = 1$ or p. Since the only divisors of p are 1 and p, it follows that $d = 1$ or p. Note that $d = p$ precisely when $p \mid a$. To illustrate this, observe that $gcd(7, 19) = 1$, while $gcd(7, 21) = 7$.

The following theorem will have major consequences. Even though it will seem obvious, we will prove it using a fact we proved earlier in this chapter.

Theorem: Let p be a prime number and suppose that $p \mid ab$. Then either $p \mid a$ or $p \mid b$.

Proof: If $p \mid a$, we are done. So we assume that p does not divide a and we must show that $p \mid b$. If p does not divide a, it follows from the previous example that $gcd(p, a) = 1$. Then there exist integers m and n such that $mp + na = 1$. Multiplying both sides of this equation by b yields $mpb + nab = b$. Now, since $p \mid ab$, it follows by Law 5, stated earlier in the chapter, that $p \mid (mpb + nab)$, implying that $p \mid b$. □

Note that this theorem is true only when p is prime. Consider the composite number 10 that divides 4×5, without dividing 4 or 5. Observe, furthermore, that the theorem can be extended to include larger products. Thus, for example, if p is a prime number and $p \mid abc$, then either $p \mid a$, $p \mid b$, or $p \mid c$. In other words, if a prime divides a product of integers, then it must divide at least one of them.

Here's another theorem whose proof is similar to the proof of the preceding one.

Theorem: Let $c \mid ab$ and let $gcd(a,c) = 1$. Then $c \mid b$.

Proof: Since $gcd(a,c) = 1$, there exist integers x and y such that $ax + cy = 1$. Multiplying by b yields $abx + cby = b$. Since $c \mid ab$, it follows that $c \mid abx + cby$, implying that $c \mid b$. $\qquad\square$

Let us now prove an interesting theorem about the Fibonacci numbers.

Theorem: Any pair of consecutive Fibonacci numbers are relatively prime, that is, $gcd(u_n, u_{n+1}) = 1$, where u_n is the nth Fibonacci number.

Proof: Recall that the Fibonacci numbers satisfy the recursive relation $u_{n+1} = u_n + u_{n-1}$, which implies that $u_{n-1} = u_{n+1} - u_n$. Now, we shall proceed with a proof by contradiction. Assume that $gcd(u_n, u_{n+1}) = d > 1$. Then $d \mid u_n$ and $d \mid u_{n+1}$, in which case $d \mid u_{n-1}$, since $u_{n-1} = u_{n+1} - u_n$. Now, observe that $u_{n-2} = u_n - u_{n-1}$, in which case $d \mid u_{n-2}$. Continuing this logic, we finally get the absurd statement $d \mid u_1$, or $d \mid 1$, and the proof is over. $\qquad\square$

We have made use, in the above proof, of Law 5, which says that $a \mid r$ and $a \mid s \Rightarrow a \mid (mr + ns)$. With $m = 1$ and $n = -1$, this says that $a \mid r$ and $a \mid s \Rightarrow a \mid (r - s)$. Here's a nice theorem.

Theorem: Given the positive integers a, b, c, and d such that $\frac{a}{b}$ and $\frac{c}{d}$ satisfy the following conditions:

1. Neither $\frac{a}{b}$ nor $\frac{c}{d}$ is an integer
2. $\frac{a}{b}$ and $\frac{c}{d}$ are reduced to lowest terms
3. $b \neq d$

then their sum, $\frac{a}{b} + \frac{c}{d}$, is not an integer.

Proof: We use a proof by contradiction. Assume that $\frac{a}{b} + \frac{c}{d} = \frac{ad + bc}{bd}$ is an integer. Then $bd \mid (ad + bc)$. Since $b \mid bd$ and $d \mid bd$, this implies that $b \mid (ad + bc)$ and $d \mid (ad + bc)$. The first of these implies that $b \mid ad$ and the second implies that $d \mid bc$. Now, Condition 2 implies that $gcd(a,b) = 1$ and $gcd(c,d) = 1$. Since $gcd(a,b) = 1$ and $b \mid ad$, it follows that $b \mid d$. By the same token, since $gcd(c,d) = 1$ and $d \mid bc$, it follows that $d \mid b$. Then $b = d$, contradicting Condition 3, and the proof is over. $\qquad\square$

DIOPHANTINE EQUATIONS

An equation such as $2x + 3y = 7$ has infinitely many solutions when x and y are real numbers. In fact, the solutions are the coordinates of any point on the straight line described by that equation. In this text, we are concerned only with *integer* solutions. You might guess that the pair of values $x = 2$ and $y = 1$ supplies such a solution. Are there others? Is there a method for finding integer solutions for equations such as this one?

An equation of the form $ax + by = c$, where a, b, and c are given integers and the solution we seek consists of a pair of integers, is called a *Diophantine equation*, in honor of the ancient Greek number theorist, Diophantus. Before we see how to solve a Diophantine equation, note that not every equation of this kind can be solved. Consider the Diophantine equation $2x + 6y = 7$. Since 2 divides the left side, we would require that $2 \mid 7$, which is not the case. Thus, there is no solution in integers. The following theorem should be obvious, in light of this example, so we shall leave the proof as an exercise at the end of the chapter.

Theorem: The Diophantine equation $ax + by = c$, where $gcd(a,b)$ does not divide c, has no solution. □

On the other hand, if $gcd(a,b) \mid c$, then the equation $ax + by = c$ has a solution. In fact, it has infinitely many solutions. Let's state this as a theorem.

Theorem: The Diophantine equation $ax + by = c$, where $gcd(a,b) \mid c$, has infinitely many solutions.

Proof: For starters, let $d = gcd(a,b)$, and assume that $d \mid c$. It then follows that we can divide a, b, and c by d, resulting in an equivalent equation in which the coefficients of x and y are relatively prime. Write this new equation as $Ax + By = C$. By a fact that we stated earlier, we can use the Euclidean algorithm to find X and Y so that $AX + BY = 1$, since A and B are relatively prime. Multiplying both sides of this equation by C yields $A(CX) + B(CY) = C$. Then $x = CX$ and $y = CY$ is a solution of the equation $Ax + By = C$, which becomes the original equation, $ax + by = c$, when we multiply both sides by d.

Let the pair of integers x_0 and y_0 (read "x naught" and "y naught") be a solution of $ax + by = c$. Then for any integer t, let

$$x = x_0 + bt$$
$$y = y_0 - at$$

We claim that the pair of integers x and y yield another solution of $ax + by = c$. To prove this, we have $a(x_0 + bt) + b(y_0 - at) = ax_0 + abt + by_0 - bat = ax_0 + by_0 = c$, since x_0 and y_0 satisfy the original equation. So we have infinitely many solutions, since any t will do.   □

In fact, *all* solutions to $ax + by = c$ are of the form $x = x_0 + bt$ and $y = y_0 - at$. To prove this, let the pair (x', y') be a solution of $ax + by = c$, where $gcd(a,b) = 1$. (Recall that this is doable when $gcd(a,b) \mid c$. Simply divide by $gcd(a,b)$.) Then $ax' + by' = c$. Now, the pair (x_0, y_0) is also a solution. Subtracting $ax_0 + by_0 = c$ yields $a(x' - x_0) + b(y' - y_0) = 0$, from which we have $a(x' - x_0) = -b(y' - y_0)$. Since b divides the right side of this equation, it follows that b divides the left side, that is, $b \mid a(x' - x_0)$. Furthermore, since $gcd(a,b) = 1$, we have $b \mid (x' - x_0)$. Then $x' - x_0 = bt$, for some integer t. Inserting this into $a(x' - x_0) = -b(y' - y_0)$ yields $abt = -b(y' - y_0)$, which becomes $at = -(y' - y_0)$. A drop of algebra then gives us $x' = x_0 + bt$ and $y' = y_0 - at$.   □

If a, b, and c are positive, there will be at most finitely many solutions where x and y are positive. This is because the line $ax + by = c$ will intercept the positive x and y axes, implying that only a finite segment of the line will be in Quadrant I. As a simple example, consider the Diophantine equation $x + y = 2$. The only solution where x and y are positive is $x = y = 1$.

Example: Find all solutions of the Diophantine equation $2x + 3y = 25$ for which x and y are both positive. We begin with $2x + 3y = 1$, since the coefficients of x and y are relatively prime. An obvious solution is $x = -1$ and $y = 1$. It follows that $x_0 = -25$ and $y_0 = 25$ is a solution to the original equation, $2x + 3y = 25$. Then all solutions are given by $x = -25 + 3t$ and $y = 25 - 2t$. To find solutions for which both of x and y are positive, we require that $-25 + 3t > 0$ and $25 - 2t > 0$. Then $\dfrac{25}{3} < t < \dfrac{25}{2}$, implying that $t = 9$, 10, 11, and 12. The positive solutions (x, y) are $(2, 7)$, $(5, 5)$, $(8, 3)$, and $(11, 1)$. Euler solved the problem as follows. Since x has the smaller coefficient, he solved for x, getting $x = \dfrac{25 - 3y}{2} = 12 - 2y + \dfrac{1 + y}{2}$. Letting $t = \dfrac{1 + y}{2}$, he obtained $y = 2t - 1$, which when substituted in the above equation yields $x = 12 - 2y + t = 12 - 2(2t - 1) + t = 14 - 3t$. So all solutions are given by

$$x = 14 - 3t$$
$$y = 2t - 1$$

Positive solutions require $14 - 3t > 0$ and $2t - 1 > 0$, or $\dfrac{1}{2} < t < \dfrac{14}{3}$. Then we have the permissible values $t = 1, 2, 3$, and 4, yielding the solutions obtained earlier.

Euler posed and solved the following problem involving a pair of Diophantine equations. A group of 30 men, women, and children spent a total of 50 dollars at an inn that charged men 3 dollars, women 2 dollars, and children 1 dollar. How many people were there in each class? Denoting the number of men, women, and children by x, y, and z, we have the two Diophantine equations

$$x + y + z = 30$$
$$3x + 2y + z = 50$$

By subtraction, we obtain $2x + y = 20$. Then $y = 20 - 2x$. Clearly, x can assume any of the values 1, 2, 3, …, 9, while y assumes the corresponding values 18, 16, 14, …, 2. The corresponding z values, 11, 12, 13, and 19, are found using $z = 30 - x - y$.

LEAST COMMON MULTIPLE

Given integers a and b, the smallest positive integer, c, such that $a \mid c$ and $b \mid c$ is called their *least common multiple* and is denoted *lcm* (a,b). Note that c is indeed a common multiple of a and b, that is, $c = ra$ and $c = sb$, where $gcd(r,s) = 1$. If $gcd(r,s) = d > 1$, then a and b would have a smaller common multiple, $\dfrac{c}{d}$. Observe that *lcm* $(a,b) \geq max\{a,b\}$, where $max\{a,b\}$ denotes the maximum of a and b. In fact, when $a \mid b$, we have $lcm(a,b) = max\{a,b\} = b$. On the other hand, we have an upper bound given by $lcm(a,b) \leq ab$. Equality occurs when $gcd(a,b) = 1$. In other words, when a and b are relatively prime, $lcm(a,b) = ab$.

Example: $lcm(15, \ 20) = 60$, $lcm(15, \ 30) = 30$, and $lcm(8, \ 9) = 8 \times 9 = 72$.

The *lcm*, c, of a and b should be familiar to you. It is the *lowest common denominator* of the (reduced) fractions $\dfrac{m}{a}$ and $\dfrac{n}{b}$. Addition of these fractions proceeds thusly $\dfrac{m}{a} + \dfrac{n}{b} = \dfrac{rm}{ra} + \dfrac{sn}{sb} = \dfrac{rm}{c} + \dfrac{sn}{c} = \dfrac{rm + sn}{c}$. Of course, when a and b are relatively prime, this becomes $\dfrac{m}{a} + \dfrac{n}{b} = \dfrac{bm}{ab} + \dfrac{an}{ab} = \dfrac{bm+an}{ab}$, since $c = lcm(a,b) = ab$.

Given a and b, since $gcd(a,b)\,|\,a$ and $a\,|\,lcm(a,b)$, it follows that gcd $(a,b)\,|\,lcm(a,b)$. Now, here are a few question that illustrate the modus operandi of the number theorist:

1. Given two distinct positive integers G and L such that $G\,|\,L$, must there exist a pair of positive integers, a and b, with $a < b$, such that $gcd(a,b) = G$ and $lcm(a,b) = L$?
2. If so, is the answer unique or are there several such pairs?
3. How many pairs are there?

For starters, note that the pair G and L answer the first question, that is, $gcd(G,L) = G$ and $lcm(G,L) = L$, since $G\,|\,L$. Question 2 is harder to answer. Sometimes, the answer is unique and sometimes it isn't. When $G = 4$ and $L = 8$, the only answer is $a = 4$ and $b = 8$. On the other hand, when $G = 10$ and $L = 300$, we have the four pairs (10, 300), (20, 150), (30, 100), and (50, 60). The third question is left as an exercise.

PRIME DECOMPOSITION

If we asked several people to factor 100, we would get a number of different answers, such as 2×50, 4×25, 5×20, and 10×10. If we stipulated that all the factors must be primes, all respondents would, except for order, give us the same answer, $2 \times 2 \times 5 \times 5$, or $2^2 \times 5^2$. This is called the *prime decomposition* of 100. We present the following important theorem.

Theorem: The prime decomposition of any positive integer greater than 1 is unique, except for the order in which the prime factors are written.

Proof: Let $n = p_1p_2...p_r = q_1q_2...q_s$ where $p_1 \le p_2 \le ... \le p_r$ and $q_1 \le q_2 \le ... \le q_s$, and all the p's and q's are prime. Then since $p_1\,|\,q_1q_2...q_s$, it follows that p_1 divides one of the q's. But this means that p_1 *equals* one of the q's. By the same logic, q_1 equals one of the p's. Since p_1 and q_1 are the smallest primes of the two prime decompositions of n, a moment's thought should convince you that they must be equal. Now, cancel them from the equation $p_1p_2...p_r = q_1q_2...q_s$, yielding the smaller equation $p_2...p_r = q_2...q_s$, which implies by the same reasoning we employed before that $p_2 = q_2$. Proceed till we exhaust the p's or the q's. At this point, if there are primes left on only one side of the equation, we get an absurd situation in which the product of those primes equals one. Then $r = s$ and the two prime decompositions are identical.   □

In light of this theorem, we see that the primes are the building blocks of the integers. In this sense, they are truly *prime*, or *first* among all numbers.

My approach to producing the prime decomposition was to do the following steps:

Produce a list of proper factors, omitting 1.

Prune that list to be just prime factors.

Slog through that list, removing a factor if it did not divide the number (more on this later) and trying it again if it did. For each actual factor, produce an element of an array holding an array 2 elements long with the factor and the exponent.

Use the array to make a display. My first approach was to produce an HTML table, but then I realized that HTML does support superscripts.

There certainly are other ways to do this. A good strategy is to divide tasks until smaller tasks.

Producing a list of proper factors is pretty straightforward. It is important (you will see this later in the definition of the tau function) that you need to be careful both in when to include one and when to include the number itself. I made use of the % operator. I also made use of the array method called push. This adds an element to an array.

As you can see from the code (end of chapter), I wrote one function, onlyPrimes, that takes all the factors and creates a new array with just the primes. This function calls a function called isPrime, which examines a single number and returns true or false.

What I call the *slog* through the list of prime factors is performed by a function, maketable, and makes use of a variable, called m in the code, that holds the current number and an array facs. Each of these variables changes during the function. If my code determines that a factor f does divide m, then the code resets m to be m/f. When the function is done checking a factor f, it removes it from facs. This is done using the shift method.

The function `displaydecomp` does the interaction. It extracts the value entered into the form by the user and displays the result in the div named place. The html `<sup>` and `</sup>` tags are used to produce the superscripting of exponents.

Example: Let p be prime. Show that $3p + 1$ is a square only when $p = 5$. If $p = 2$ or 3, we obtain the nonsquares 7 and 10, respectively. So assume $p > 3$. Now, if $3p + 1 = n^2$, then $3p = n^2 - 1 = (n - 1)(n + 1)$. By the above theorem, $3p$ is the unique prime factorization of $n^2 - 1$, from which $n - 1 = 3$ and $n + 1 = p$. Then $n = 4$ and, therefore, $p = 5$.

Example: Let n be a positive integer. Show that the smallest nontrivial divisor d of n is a prime number. (Here, *nontrivial* means that $d > 1$.) If d is composite, it has a prime divisor $p < d$. Since $p \mid d$ and $d \mid n$, we have $p \mid n$, contradicting the assumption that d is the smallest nontrivial divisor of n.

Recall from Chapter 2 that to determine whether a given positive integer n is a prime, it suffices to check whether n is divisible by any of the numbers less than or equal to $\sqrt{n}$. By the above example, the observation can now be improved. We have the following remark.

Remark: To determine whether a positive integer n is a prime, it suffices to check whether n is divisible by any of the *prime* numbers less than or equal to $\sqrt{n}$.

Example: Verify that 101 is prime. The primes less than $\sqrt{101}$ are 2, 3, 5, and 7. Since neither of these primes divides 101, the claim is verified. 101 is indeed prime.

SEMIPRIME NUMBERS

A positive integer n is called *semiprime* if $n = pq$ where p and q are distinct primes. Thus, $6 = 2 \times 3$, $10 = 2 \times 5$, and $14 = 2 \times 7$ are the first three semiprimes. Here is an interesting theorem about semiprimes.

Theorem: If the smallest prime divisor, p, of n satisfies $p > n^{\frac{1}{3}}$, then n is semiprime.

Proof: Since $p \mid n$, it follows that $n = pd$. The theorem will follow if we show that d is prime. Observe that $d < n^{\frac{2}{3}}$. (If $d \geq n^{\frac{2}{3}}$, we would have $n = pd > n^{\frac{1}{3}} \times n^{\frac{2}{3}} = n$, which becomes $n > n$, which is absurd.) Now, if d is composite, then d has a prime divisor q such that $q \leq \sqrt{d}$. Since $d < n^{\frac{2}{3}}$, we have $q \leq \sqrt{d} < n^{\frac{1}{3}}$. The fact that q is a prime divisor of n contradicts the assumption that the smallest prime divisor, p, of n satisfies $p > n^{\frac{1}{3}}$. $\qquad\square$

Example: The smallest prime divisor, 2, of 6 exceeds $6^{\frac{1}{3}}$ (since $2^3 > 6$), implying that 6 is semiprime.

The condition of the theorem need not be present when n is semiprime as is shown by the fact that the smallest prime divisor, 2, of the semiprime 10 does not exceed $10^{\frac{1}{3}}$. It is not an "if and only if" theorem. The next theorem will be proven using prime decomposition.

Theorem: Let a and b be positive integers. Then $gcd(a,b) \times lcm(a,b) = ab$.

Proof: Let $a = p_1^{e_1} p_2^{e_2} \ldots p_r^{e_r}$ and let $b = p_1^{f_1} p_2^{f_2} \ldots p_r^{f_r}$. This is possible, since we can combine the primes of a and b into the sequence of primes, $p_1, p_2, \ldots, p_r$. The primes that occur in a but do not occur in b receive an exponent of 0 in b, and the same goes for the primes that occur in b but not in a. (If $a = 10$ and $b = 18$, e.g., then the primes are 2, 3, and 5. Then we write $10 = 2^1 3^0 5^1$ and $18 = 2^1 3^2 5^0$.)

Now, let $m_1 = min\{e_1, f_1\}$, $m_2 = min\{e_2, f_2\}$, ..., $m_r = min\{e_r, f_r\}$, and let $M_1 = max\{e_1, f_1\}$, $M_2 = max\{e_2, f_2\}$, ..., $M_r = max\{e_r, f_r\}$. Then $gcd(a,b) = p_1^{m_1} p_2^{m_2} \ldots p_r^{m_r}$, and $lcm(a,b) = p_1^{M_1} p_2^{M_2} \ldots p_r^{M_r}$.

It follows that $gcd(a,b) \times lcm(a,b) = p_1^{m_1 + M_1} p_2^{m_2 + M_2} \ldots p_r^{m_r + M_r}$. Now, we need the fact that given x and y, then $min(x,y) + max(x,y) = x + y$, from which it follows that for each $i = 1, 2, \ldots, r$, we have $m_i + M_i = e_i + f_i$. Then the product $p_1^{m_1 + M_1} p_2^{m_2 + M_2} \ldots p_r^{m_r + M_r} = p_1^{e_1 + f_1} p_2^{e_2 + f_2} \ldots p_r^{e_r + f_r} = p_1^{e_1} p_2^{e_2} \ldots p_r^{e_r} \times p_1^{f_1} p_2^{f_2} \ldots p_r^{f_r} = ab$. $\qquad\square$

WHEN IS A NUMBER AN *m*TH POWER?

The next theorem tells us when a given number is a square, a cube, or, more generally, a *power*. In other words, when is a given number, n, an *m*th power? In still other words, when does n equal c^m?

Theorem: Let $n = p_1^{e_1} p_2^{e_2} \ldots p_r^{e_r}$. Then n is an mth power if and only if $m \mid e_i$ for each $i \leq r$.

Proof: An "if and only if" proof requires two parts. Firstly, we assume that $m \mid e_i$ for each $i = 1, 2, \ldots, r$ and show that n is an mth power. Now, $m \mid e_i$ implies that $e_i = mf_i$ for each $i = 1, 2, \ldots, r$. (In other words, $e_1 = mf_1$, $e_2 = mf_2$, etc.) Then $n = p_1^{e_1} p_2^{e_2} \ldots p_r^{e_r} = \left(p_1^{f_1} p_2^{f_2} \ldots p_r^{f_r} \right)^m$, proving that n is an mth power. For the second part of the proof, we assume that n is an mth power and we show that $m \mid e_i$ for each $i = 1, 2, \ldots, r$. To do this, let $n = c^m$. Using the prime decomposition theorem, let $c = p_1^{f_1} p_2^{f_2} \ldots p_r^{f_r}$. Then $n = c^m = \left(p_1^{f_1} p_2^{f_2} \ldots p_r^{f_r} \right)^m = p_1^{mf_1} p_2^{mf_2} \ldots p_r^{mf_r}$, in which case each power is indeed a multiple of m.   □

The statement $a \mid b$ has a simple interpretation using prime decomposition. Whenever a prime p occurs in the prime decomposition of a with exponent r, it must occur in the prime decomposition of b with exponent s, where $s \geq r$. Thus, $20 \mid 600$, since $20 = 2^2 5^1$, and $600 = 2^3 3^1 5^2$. Using this observation, the proof of the next theorem will be easy.

Theorem: Let the relatively prime pair of numbers a and b divide c. Then $ab \mid c$.

Proof: Since a and b divide c, it follows that c contains all the prime divisors of a and b and the exponents of these primes in c are greater than or equal to the corresponding exponents in a and b. Since the prime decompositions of a and b have no common primes, the conclusion follows.   □

Observe that the above theorem is not valid when a and b are not relatively prime. Consider the fact that 6 and 10 divide 30, while their product, 60, does not. The problem here is that 6 and 10 both contain the factor 2, which 30 also contains, while the product 6×10 contains the factor 2^2, which 30 does not contain.

Example: Show that $6 \mid n(n + 1)(2n + 1)$. It suffices to show that $2 \mid n(n + 1)(2n + 1)$ and that $3 \mid n(n + 1)(2n + 1)$. The first statement is easy. Since n and $n + 1$ are consecutive, one of them must be even, implying that $2 \mid n(n + 1)$. Then most certainly $2 \mid n(n + 1)(2n + 1)$. The second statement requires three cases, since n must have exactly one of the forms $3k$, $3k + 1$, or $3k + 2$. If $n = 3k$, we are done, since $3 \mid n$. If $n = 3k + 1$, then $2n + 1 = 6k + 3 = 3(2k + 1)$, so $3 \mid (2n + 1)$, and we are done. Finally, if $n = 3k + 2$, then $n + 1 = 3k + 3 = 3(k + 1)$, implying that $3 \mid (n + 1)$.

TWIN PRIMES

While odd primes cannot be consecutive, here and there, we find pairs of primes, such as 17 and 19, such that their difference is 2. We call them *twin primes*. The first few are 3 and 5, 5 and 7, 11 and 13, 17 and 19, and 29 and 31. It is currently not known if there are infinitely many pairs of twin primes. Nevertheless, here is a nice theorem concerning them.

Theorem: Let p and $p + 2$ be twin primes, such that $p \geq 5$. Then their sum is a multiple of 12, that is, $12 \mid 2p + 2$.

Proof: All numbers have one of the forms $6k$, $6k \pm 1$, $6k \pm 2$, or $6k + 3$. (Interesting! No number can be more than 3 away from a multiple of 6.) A prime p such that $p \geq 5$ can only have the form $6k \pm 1$, since the other forms are divisible by 2 or 3. Since the difference of the two primes p and $p + 2$ is 2, we must have $p = 6k - 1$ and $p + 2 = 6k + 1$. Then their sum $2p + 2 = 12k$. □

Observe that all positive integers greater than 11 can be written as $12k + r$, where $k \geq 1$ and $r = 0, 1, 2, \ldots, 11$. If p is prime, we can write $p = 12k + r$, but r is now restricted to the values 1, 5, 7, and 11. If $r = 3$, for example, we have $p = 12k + 3 = 3(4k + 1)$, a contradiction.

Theorem: Let $p \geq 5$ be a prime. Then $24 \mid p^2 - 1$.

Proof: Note that the theorem is true for the primes 5, 7, and 11. Now, by the observation prior to the theorem, all primes greater than 11 can be written $12k + r$, where $k \geq 1$ and $r = 1, 5, 7,$ or 11. Setting $p = 12k + r$, we get

$$p^2 - 1 = (12k + r)^2 - 1 = 144k^2 + 24kr + r^2 - 1 = 24(6k^2 + kr) + (r^2 - 1)$$

Now, $r^2 - 1 = 0, 24, 48,$ or 120, each of which is divisible by 24. □

FERMAT PRIMES

The great seventeenth-century number theorist Pierre Fermat (1601–1665) conjectured that $2^{2^n} + 1$ is always prime. This is true when $n = 0, 1, 2, 3,$ and 4. In the following century, Euler found that when $n = 5$, Fermat's expression is composite, by showing that $641 \mid 2^{32} + 1$. While this is easy to do today with the help of a computer, here is a number theoretic argument.

Let $x = 2^7 = 128$ and let $y = 5$. Then $1 + xy = 1 + 5 \times 128 = 641$. We also have $1 + xy - y^4 = 1 + y(x - y^3) = 1 + 5 \times 3 = 16 = 2^4$. Finally, $2^{32} + 1 = 2^4 2^{28} + 1 = 2^4 (2^7)^4 + 1 = (1 + xy - y^4)x^4 + 1 = (1 + xy)x^4 + (1 - x^4 y^4) = (1 + xy)x^4 + (1 - x^2 y^2)(1 + x^2 y^2) = (1 + xy)[x^4 + (1 - xy)(1 + x^2 y^2)] = 641[x^4 + (1 - xy)(1 + x^2 y^2)]$, and we are done. To date, it is not known whether there are any more *Fermat primes*, that is, primes of the form $2^{2^n} + 1$. The number $2^{2^7} - 1$ wasn't factored until 1971. We denote the nth Fermat number, $2^{2^n} + 1$, by F_n. As of this writing, no Fermat primes have been found beyond F_4. Before we leave the topic of Fermat numbers, here is a theorem with a surprising consequence.

Theorem: The Fermat numbers F_n and F_m where $m \neq n$ are relatively prime.

Proof: Let $n > m$ and let $d = gcd(F_n, F_m)$. Since F_n and F_m are odd, it follows that d is odd. We will show that $d = 1$. Let $x = 2^{2^m}$ and let $a = 2^{n-m}$. Then $F_n - 2 = 2^{2^n} - 1 = x^a - 1$. Since a is even, $x = -1$ is a root of the equation $x^a - 1 = 0$, implying that $x + 1 \mid x^a - 1$. Then $F_m \mid F_n - 2$. Since $d \mid F_m$, it follows that $d \mid F_n - 2$. Now, since, in addition, $d \mid F_n$, we have $d \mid 2$, implying that $d = 1$ since d is odd, and the proof is over. $\square$

It follows that the nth Fermat number, F_n, contains at least one prime divisor that does not divide any of the Fermat numbers, F_m, where $m < n$. So there are infinitely many primes!

ODD PRIMES ARE DIFFERENCES OF SQUARES

$\lfloor x \rfloor$ and $\lceil x \rceil$, where x is a nonnegative real number, mean "round down x" and "round up x," respectively. Thus, $\lfloor 3.14 \rfloor = 3$, while $\lceil 3.14 \rceil = 4$. Similarly, $\lfloor \pi \rfloor = 3$ and $\lceil \pi \rceil = 4$. When x is an integer, $\lfloor x \rfloor = \lceil x \rceil = x$. Using this notation, here's an interesting theorem.

Theorem: Let p be an odd prime. Then p can be written *uniquely* as the difference of squares.

Proof: Let $p = a^2 - b^2$. Then $p = (a - b)(a + b)$. Since $a + b > a - b$ and since p can only be factored as $1 \times p$, we *must* have

$$1 = a - b$$
$$p = a + b$$

which, upon adding and subtracting these equations and dividing by 2, yield

$$a = \frac{p+1}{2}$$

$$b = \frac{p-1}{2}$$

Since p is odd, it follows that $p+1$ and $p-1$ are even, in which case we see that a and b are integers. In fact, $a = \left\lceil \frac{p}{2} \right\rceil$ and $b = \left\lfloor \frac{p}{2} \right\rfloor$, so $p = a^2 - b^2 = \left\lceil \frac{p}{2} \right\rceil^2 - \left\lfloor \frac{p}{2} \right\rfloor^2$, and the proof is complete. □

Example: Write 19 as the difference of two squares. By the above proof, $a = \lceil 9.5 \rceil = 10$, and $b = \lfloor 9.5 \rfloor = 9$. Indeed, $19 = 100 - 81$.

Example: Show that the composite number 15 can be written in two different ways as the difference of squares. By the method of the above proof, $a = \lceil 7.5 \rceil = 8$, and $b = \lfloor 7.5 \rfloor = 7$, implying that $15 = 64 - 49$. On the other hand, the equation $15 = (a-b)(a+b)$ also yields the system of equations

$$3 = a - b$$
$$5 = a + b$$

So $a = 4$ and $b = 1$, and $15 = 16 - 1$.

WHEN IS *n* A LINEAR COMBINATION OF *a* AND *b*?

Example: Given the positive numbers a and b such that $\gcd(a,b) = 1$, show that there is a positive number, N, such that for any $n > N$, it is possible to write n as a sum of *positive* multiples of a and b, that is, $n = ra + sb$, where r and s are positive integers. Note first that since $\gcd(a,b) = 1$, there exist integers x and y such that $ax + by = 1$, where x is positive and y is negative. (If x and y are both negative, it would follow that $ax + by$ is negative. If x and y are both positive, it would follow that $ax + by$ is greater than 1.) Let's change the plus sign into a minus sign and let y be positive, that is, we have $ax - by = 1$, where x and y are positive.

All integers can be written $mb + r$, where $0 \le r < b$. Assume that the N we are looking for is of the form Mb. Consider all numbers, n, of the form $mb + r$, where $m \ge M$ and $0 \le r < b$. Now, $n = mb + r = mb + r \times 1 = mb + r$ $(ax - by) = rax + (m - ry)b$.

Clearly, $m - ry$ is the problem here. We must show that we can make this number positive. We have the following inequality: $m - ry \ge M - ry \ge M - (b - 1)y$. It suffices, therefore, to choose M so that $M > (b - 1)$ y. Finally, since $N = Mb$, we can let $N = b(b - 1)y$.

Note that if $gcd(a,b) = d > 1$, no N exists, since $d \mid ra + sb$, for all r and s.

PRIME DECOMPOSITION OF $n!$

Let's obtain the prime decomposition of $n!$. Now, p is a prime factor of $n!$ if and only if $p \le n$. Call these primes p_1, p_2, ..., p_r, and let their exponents be e_1, e_2, ..., e_r. So $n = p_1^{e_1} p_2^{e_2} ... p_r^{e_r}$.

To determine e_i, first find the number of integers up to n containing p_i as a factor. This number is $\left\lfloor \dfrac{n}{p_i} \right\rfloor$. For example, the number of integers up to 65 that contain 3 as a factor is $\left\lfloor \dfrac{65}{3} \right\rfloor = 21$, implying that $3^{21} \mid 65!$. But this number is too small, since numbers such as 9 and 18 contain more than one "3" as a factor. Thus, we must add $\left\lfloor \dfrac{65}{3^2} \right\rfloor = \left\lfloor \dfrac{65}{9} \right\rfloor = 7$ to the previously obtained 13. In general, we must add $\left\lfloor \dfrac{n}{(p_i)^2} \right\rfloor$ to $\left\lfloor \dfrac{n}{p_i} \right\rfloor$. Returning to the specific example, 65!, since 27 and 54 contains three "3s" as factors, we must add on $\left\lfloor \dfrac{65}{3^3} \right\rfloor = \left\lfloor \dfrac{65}{27} \right\rfloor = 2$, obtaining the exponent $21 + 7 + 2 = 30$ of 3 in the prime decomposition of 65!. What would happen if we continued this example and divided by the next power of 3, that is, 3^4 or 81? Since we are rounding down, the answer is "no harm done" since $\left\lfloor \dfrac{65}{81} \right\rfloor = 0$.

In fact, the exponent of 3 in the prime decomposition of 65! may be written as the infinite series $\displaystyle\sum_{k=1}^{\infty} \left\lfloor \dfrac{65}{3^k} \right\rfloor = \left\lfloor \dfrac{65}{3} \right\rfloor + \left\lfloor \dfrac{65}{3^2} \right\rfloor + \left\lfloor \dfrac{65}{3^3} \right\rfloor + \cdots$. In the spirit of this idea, we have the following.

Theorem: The prime decomposition of $n!$ is given by $p_1{}^{e_1}p_2{}^{e_2}...p_r{}^{e_r}$, where the primes are precisely those satisfying $p_i \leq n$, and the exponent e_i of p_i is given by

$$e_i = \sum_{k=1}^{\infty} \left\lfloor \frac{n}{p_i{}^k} \right\rfloor = \left\lfloor \frac{n}{p_i} \right\rfloor + \left\lfloor \frac{n}{p_i{}^2} \right\rfloor + \left\lfloor \frac{n}{p_i{}^3} \right\rfloor + \cdots \qquad \square$$

Example: Find the prime decomposition of 10!. The primes less than 10 are 2, 3, 5, and 7. Their exponents are given by $\left\lfloor \frac{10}{2} \right\rfloor + \left\lfloor \frac{10}{4} \right\rfloor + \left\lfloor \frac{10}{8} \right\rfloor = 8$, $\left\lfloor \frac{10}{3} \right\rfloor + \left\lfloor \frac{10}{9} \right\rfloor = 4$, $\left\lfloor \frac{10}{5} \right\rfloor = 2$, and $\left\lfloor \frac{10}{7} \right\rfloor = 1$. So $10! = 2^8 3^4 5^2 7^1 = 256 \times 81 \times 25 \times 7 = 3{,}628{,}800$.

NO NONCONSTANT POLYNOMIAL WITH INTEGER COEFFICIENTS ASSUMES ONLY PRIME VALUES

Recall that we proved in the previous chapter that no nonconstant polynomial $f(x)$ with integer coefficients assumes only prime values when $x = 1$, 2, 3, ... We prove this using divisibility.

It is easy to verify the algebraic identity $a^n - b^n = (a-b)$ $(a^{n-1} + a^{n-2}b + \cdots + ab^{n-2} + b^{n-1})$. When $n = 3$, by the way, it tells us how to factor the difference of two cubes:

$$a^3 - b^3 = (a-b)(a^2 + ab + b^2),$$

a fact not as well remembered as the way to factor the difference of two squares. It follows that $a - b \mid a^n - b^n$, for $n = 1$, 2, 3, ... Consider $f(x) = c_k x^k + c_{k-1}x^{k-1} + c_{k-2}x^{k-2} + \cdots + c_2 x^2 + c_1 x + c_0$, where the coefficients are integers. We find that

$$f(b) - f(a) = \left(c_k b^k + c_{k-1}b^{k-1} + \cdots + c_1 b + c_0\right) - \left(c_k a^k + c_{k-1}a^{k-1} + \cdots + c_1 a + c_0\right)$$

So

$$f(b) - f(a) = c_k\left(b^k - a^k\right) + c_{k-1}\left(b^{k-1} - a^{k-1}\right) + c_{k-2}\left(b^{k-2} - a^{k-2}\right) + \cdots + c_2\left(b^2 - a^2\right) + c_1(b-a)$$

Now, since each term on the right contains a factor of the form $a^n - b^n$ and since we know that $a - b \mid a^n - b^n$, for $n = 1, 2, 3, \ldots, k$, we have the proven the following theorem.

Theorem: Let $f(x) = c_k x^k + c_{k-1} x^{k-1} + c_{k-2} x^{k-2} + \cdots + c_2 x^2 + c_1 x + c_0$, where the coefficients are integers, and let a and b be given integers. Then $a - b \mid f(a) - f(b)$.   □

Alternatively, we have $f(a) - f(b) = K(a - b)$, where K is an integer. We are ready to prove the following theorem.

Theorem: No polynomial $f(x)$ with integer coefficients assumes only prime values when $x = 1, 2, 3, \ldots$ (unless the polynomial is a constant polynomial such as $f(x) = 7$).

Proof: Let n be a positive integer for which $f(n)$ is prime. (If this isn't possible, the proof is over.) Let m be any positive integer and let

$$a = mf(n) + n$$
$$b = n$$

Then by the preceding theorem, $a - b \mid f(a) - f(b)$, or $mf(n) \mid f[mf(n) + n] - f(n)$, which can be restated as $f[mf(n) + n] - f(n) = Kmf(n)$, where K is an integer. Writing this as

$$f[mf(n) + n] = Kmf(n) + f(n) = f(n)[Km + 1]$$

we see that $f(x)$ is composite for the infinitely many values $x = mf(n) + n$.   □

EXERCISES

An asterisk (∗) indicates that the exercise can be developed into a research project.

1   Write a program to find the greatest common divisor of a given pair of positive integers. If the answer is 1, the program should state that the two numbers are relatively prime. Then find *gcd* (80,540), *gcd*(18,600), and *gcd*(105,350). *At the end of the chapter, you can find a program that determines the GCD of the two numbers entered into the form by the user. See if you can add the*

message stating that two numbers are relatively prime when that is the case.

2   Verify that $6 \mid n^3 - n$ for $n \leq 1000$. Then prove it is true for all n.

3   Repeat the previous exercise for the statement $30 \mid n^5 - n$.

4   *If m and n are odd, show that $2 \mid m^2 + n^2$ but 4 does not divide $m^2 + n^2$ for all m and n up to 1000.

5   *Show that a number of the form $n(n+1)(n+2)$ is not a square for $n \leq 1000$.

6   *Show that a number of the form $n(n+1)(n+2)$ is not a cube for $n \leq 1000$.

7   Given the *distinct* (unequal) positive integers a and b less than 100, show that $\frac{1}{a} + \frac{1}{b}$ is not an integer.

8   *Show that $4 \mid 3^n + 1$ when n is odd and less than 100. Try to prove this for all n.

9   Prove that $5 \mid u_{5k}$, that is, show that every fifth Fibonacci number is a multiple of 5.

10 *For which positive even values of n up to 100 does $2n + 1$ divide $n^4 + n^2$?

11 Find solutions of the Diophantine equations $6x + 9y = 21$, $5x + 7y = 64$, and $3x + 8y = 9$.

12 Show that $2^{2^n} + 1$ is prime for $n = 0, 1, 2, 3$, and 4.

13 Use one of the rounding functions to find $f(n)$ if the values of f when $n = 1, 2, 3, 4, 5, 6, 7, 8$, and 9 are 1, 1, 1, 2, 2, 2, 3, 3, and 3, respectively.

14 How large can $\lfloor 10x \rfloor - 10\lfloor x \rfloor$ get? Assume that x is a positive real number.

15 Write n^3 as the difference of two squares for all $n \leq 100$.

16 Show that $2^r + 1 \mid 2^{3r} + 1$ for all $r \leq 100$.

17 Show that 4 does not divide $n^2 + 1$ for all $n \leq 100$. Then show that this is true for all n.

18 Find the prime decomposition of 30!.

19 Let p be a prime number less than 101. Show that $p \mid \binom{p}{k}$ where $1 \leq k \leq p - 1$.

20 Write a program that determines whether a given positive integer n is prime by checking whether it is divisible by any of the primes less than or equal to $\sqrt{n}$.

21 Write a program to find all the ways in which a given positive integer may be written as a product of two divisors. The program should also state the number of ways in which this can be done.

22 Verify *Bertrand's postulate* (that for $n > 1$, there is a prime number between n and $2n$) for all n such that $1 < n \le 1000$.

23 *An arithmetic progression of length k is a sequence of the form a, $a + d, a + 2d, a + 3d, \ldots, a + (k-1)d$. The sequence 3, 7, 11, 15, 19, for example, is an arithmetic progression of length 5. Write a program that produces arithmetic progressions of length k consisting of primes. The sequence 5, 11, 17, 23, 29, for example, is an arithmetic progression of length 5 consisting of primes. Note that $d = 6$ is even. Explain why d must be even in an arithmetic progression of primes.

24 *Find 50 primes in the arithmetic progression 113, 213, 313, 413, … (in fact, there are infinitely many primes in this progression).

Greatest Common Divisor

```
<!DOCTYPE HTML>
<html>
<meta charset="UTF-8">
<head>
<title>GCD</title>
<script>
var placeref;
function find() {
placeref = document.getElementById("place");
placeref.innerHTML = "";

gcd(parseInt(document.f.numA.value),parseInt
(document.f.numB.value));
return false;
}
function gcd(a,b) {
  var messages;
  var holder;
  var r;
  if (b < a) {
```

```
  //switch
  holder = a;
  a = b;
  b = holder;
  }
  messages = placeref.innerHTML;
  messages += "<br/> Searching for GCD of "+String
(a)+" and "+String(b)+".<br/>";
  placeref.innerHTML = messages;
  r = b % a;
  if (r==0){
    messages+= "GCD is " + String(a);
    placeref.innerHTML = messages;
    return false;
  }

  else {
    return gcd(r,a);

  }
}
</script>
</head>
<body>
Enter two numbers <br/>
<form name="f" onsubmit="return find();">
A: <input type="number" name="numA" value=""/> 
 B: <input type="number" name="numB" value=""/>
<br/>
  <input type="submit" value="Enter"/>
</form>
<p id="place">
  Messages will go here.
</p>
</body>
</html>
```

Prime Decomposition

```
<!DOCTYPE HTML>
<html>
<meta charset="UTF-8">
```

```
<head>
<title>Primary Decomposition</title>
<script>

var table = [] ;
//each element will be an array of length 2

function produceFactors (n) {
  var ans = [] ;  //start off with empty array
  //notice start with 2, not 1
  for (var i=2;i<n;i++) {
      if ((n%i)==0) {
          ans.push(i) ;


      }
  }


  return ans ;
}
function onlyprimes (factors) {
  var ans = [] ;
  for (var i=0;i<factors.length;i++) {
      if (isPrime(factors[i])) {
        ans.push(factors[i]) ;
      }
  }
  return ans ;


}
function makeTable (m,facs) {
    while (facs.length>0) {
      f = facs[0] ;
//always using first, that is, 0th element
      if ((m % f )>0) {
  //f, the ith element of facs does NOT divide m
          facs.shift () ;  //remove this factor
      }
      else {
        m = m / f ;
        e=1;
        //need to see if exponent is more than 1
```

```
            while (0 == (m%f ) ) {
                m = m /f;
                e++;
            }
            table.push ([f,e]);
            facs.shift();
        }

    }
    return;
}

function isPrime(n) {
  var lim = Math.sqrt(n);
  for (var i=2;i<=lim;i++) {
     if (0 == n % i) {
        return false;
     }
  }
  return true;

}
function displaydecomp() {
  table = [];  //reset table to empty array
  placeref = document.getElementById("place");
  n = parseInt(document.f.num.value);
  factors = produceFactors(n);
  pfactors = onlyprimes(factors);
  makeTable(n, pfactors);
// this function will add to table
 messages = "The prime decomposition is <br/>";

  for (var i=0;i<table.length;i++) {

messages+=String(table[i][0])+"<sup>"+String
(table[i][1])+"</sup>";
  }

  placeref.innerHTML = messages;
  return false;
}
</script>
</head>
```

```
<body>
Enter number to see its prime decomposition:
<form name="f" onSubmit="return
displaydecomp();">
 <input type="number" name="num" value=""/>
 <input type="submit" value="Enter integer"/>
</form>
<div id="place">
Answer will go here.
</div>
</body>
</html>
```

5

MODULAR ARITHMETIC

Modular arithmetic sheds light on the relation of integers to their remainders when they are divided by a given positive integer. It is useful in both number theory and computer science.

CONGRUENCE CLASSES MOD k

The integers behave in a *periodic* way when we examine their expressions as one of the forms, for example, $3n$, $3n + 1$, and $3n + 2$. Every third integer is a multiple of 3 and can therefore be written $3n$ for some integer n. Thus, $3 = 3 \times 1$, $6 = 3 \times 2$, $9 = 3 \times 3$, $12 = 3 \times 4$, and so on. On the other hand, the numbers 4, 7, 10, 13, etc. can be written $3n + 1$, since they are one more than some multiple of 3. Similarly, 5, 8, 11, 14, etc. can be written $3n + 2$. Thus, the sequence of consecutive numbers 3, 4, 5 yield numbers of the form $3n$, $3n + 1$, and $3n + 2$. The next number, 6, signifies that the pattern repeats. In fact, the next three numbers, 6, 7, 8, mimic their three predecessors perfectly, following the same format as $3n$, $3n + 1$, $3n + 2$. Of course, the n changes from 1 to 2, but the pattern persists. Every third number is of the same format.

Elementary Number Theory with Programming, First Edition. Marty Lewinter and Jeanine Meyer.
© 2016 John Wiley & Sons, Inc. Published 2016 by John Wiley & Sons, Inc.

There are three *classes* of numbers: $3n$, $3n + 1$, and $3n + 2$. All numbers belong to one of these mutually exclusive and jointly exhaustive classes. Even 0 is a member of one of the classes, namely, the $3n$ class. Negative numbers are included, as well, since n may be negative.

The entire analysis can be repeated for the coefficient 5 or any other coefficient. Every integer is in one of the forms $5n$, $5n + 1$, $5n + 2$, $5n + 3$, and $5n + 4$. Of course, if we continued this consecutive sequence to $5n + 5$, we would needlessly be introducing an extra class, since $5n + 5 = 5(n + 1)$, which is in the same class as $5n$. One could let $m = n + 1$, in which case $5n + 5 = 5m$. There's nothing sacred about n.

Notice there are five classes when we use the coefficient 5 and, more generally, there are k classes when we use the coefficient k. Instead of the word *coefficient*, mathematicians refer to the *modulus k*. The k classes are called *congruence classes mod k*, and the whole subject is called *modular arithmetic*. Two numbers, a and b, in the same class are called *congruent mod k*. This is written $a = b$ (*mod k*). If a and b belong to different classes, they are called *incongruent mod k*. This is written $a \neq b$ (*mod k*).

The chart below, employing the modulus 4, demonstrates the periodic nature of modular arithmetic. The three dots in the first and last box indicate that the chart extends in both directions forever.

...	−3	−2	−1
0	**1**	**2**	**3**
4	5	6	7
8	9	10	11
12	13	14	15
16	17	18	19
20	21	22	23
24	25	26	...

Every number in the first column is of the form $4n$, that is, it is a multiple of 4. Every number in the second column is of the form $4n + 1$, that is, one more than a multiple of 4, and so on. The bold numbers 0, 1, 2, and 3 are called *least residues*. Each is the remainder when any positive number in its column is divided by 4.

Two numbers are in the same column if and only if their difference is a multiple of 4. Thus, 21 and 9 are in the same column, since $21 - 9 = 12$, which is a multiple of 4. Alternatively, $21 = 9$ (*mod 4*) since these numbers are in the same *congruence class*—namely, the class whose least residue is 1. Observe that this is the remainder when either number is divided by 4.

If we employ the modulus 2, we get a chart with only two columns. The least residues are 0 and 1. These are the remainders, of course, when we divide numbers by 2. Notice that $n = 0$ (*mod* 2) when n is even and $n = 1$ (*mod* 2) when n is odd. The *parity* of a number (whether it is odd or even) is the congruence class (*mod* 2) to which it belongs.

LAWS OF MODULAR ARITHMETIC

The following five statements are different ways of saying the same thing:

1. $a = b$ (*mod n*).
2. $a - b$ is a multiple of n, that is, $n \mid a - b$, or n *divides* $a - b$.
3. a and b are in the same column of a modular chart for n, that is, a and b are in the same congruence class *mod n*.
4. a and b have the same remainder when they are divided by n.
5. a and b have the same form, that is, one is $kn + r$ and the other is $jn + r$, where r satisfies $0 \le r \le n - 1$.

It is quite interesting to note that congruence equations have most of the properties of ordinary equations. Here are a few of them:

1. If $a = b$ (*mod n*), then $b = a$ (*mod n*).
2. If $a = b$ (*mod n*) and $b = c$ (*mod n*), then $a = c$ (*mod n*).
3. If $a = b$ (*mod n*), then $a \pm c = b \pm c$ (*mod n*).
4. If $a = b$ (*mod n*) and $c = d$ (*mod n*), then $a \pm c = b \pm d$ (*mod n*).
5. If $a = b$ (*mod n*), then $ac = bc$ (*mod n*).
6. If $a = b$ (*mod n*) and $c = d$ (*mod n*), then $ac = bd$ (*mod n*).
7. If $a = b$ (*mod n*), then $a^c = b^c$ (*mod n*).

Note that $ac = bc$ (*mod n*) does not permit us to conclude that $a = b$ (*mod n*). As an illustration, canceling a 6 in the congruence equation $12 = 18$ (*mod* 6) yields the false equation $2 = 3$ (*mod* 6). When can one cancel with impunity? When the factor and the modulus are *relatively prime*. Consider the equation $20 = 50$ (*mod* 6). Upon canceling the factor 5, we obtain the correct equation $4 = 10$ (*mod* 6). Of course, 5 and 6 are relatively prime, so the cancelation was assured to be legitimate. Another situation that permits cancelation is $ac = bc$ (*mod nc*). This means that $nc \mid ac - bc$, or $nc \mid c(a - b)$, implying that $n \mid a - b$, or $a = b$

(*mod n*). The price we pay in this case, however, is a change in the modulus from *cn* to *n*.

Example: Note that ordinary equations must remain true when we suddenly append the expression (*mod n*) to it, but, obviously, not vice versa. Since $n = 0, 1, 2$, or 3 (*mod* 4), it follows, using the laws of modular arithmetic, that $n^2 = 0$ or 1 (*mod* 4). This is because $2^2 = 4 = 0$ (*mod* 4) and $3^2 = 9 = 1$ (*mod* 4). The first few squares are in the 0 class or the 1 class *mod* 4. In fact, the squares alternate between these two classes: $1 = 1$ (*mod* 4), $4 = 0$ (*mod* 4), $9 = 1$ (*mod* 4), $16 = 0$ (*mod* 4), $25 = 1$ (*mod* 4), $36 = 0$ (*mod* 4), $49 = 1$ (*mod* 4), and so on. We can easily prove, then, that numbers of the form $4n + 3$ are never expressible as the sum of two squares. Let $s = 4n + 3$, and assume that $s = a^2 + b^2$, that is, assume that *s* is the sum of two squares. Then when we convert this equation to a congruence *mod* 4, it says that 3 is the sum of two numbers from the set $\{0, 1\}$, which is absurd. So the desired conclusion follows.

Determining the truth or falseness of an equation $a = b$ (*mod n*) is straightforward given the presence of the % operator in JavaScript (and similar operators in other programming languages). What we check is if $(a - b)$ is equal to zero *mod n*. To put it another way, our code computes:

```
r =   (a-b) % n;
```

and checks if r is equal to zero:

```
if (r==0) ...
```

You can examine the program at the end of the chapter. Notice that my program has a form for the user to enter values for a, b and n and I use a `div` element in the code to display messages as I have done before.

The earlier comments about the "form" of a number may be modified to take into account the following alternate classification. All numbers may be written as $3n$, $3n + 1$, or $3n - 1$. After all, a number of the form $3n + 2$, which is *two more* than a multiple of three, is *one less* than the next multiple of three, that is, it is of the form $3n - 1$. Similarly, all numbers are in one of the five classes $5n$, $5n \pm 1$, and $5n \pm 2$. In other words, no number is more than two away from a multiple of 5. One sees similarly that no number is more than three away from a multiple of 7. More generally, no number is more than k away from a multiple of the odd number $2k + 1$.

Example: What is the remainder when 10^{100} is divided by 9? Since $10 = 1$ $(mod\,9)$, we have $10^{100} = 1^{100}\,(mod\,9) = 1\,(mod\,9)$. Thus, the remainder is 1.

Example: Show that $1! + 2! + 3! + \cdots + 1000! = 3(mod\,6)$. Well, $3! = 0$ $(mod\ 6)$ and the same is true for $n!$, when $n > 3$. Then $1! + 2! + 3! + \cdots + 1000! = (1! + 2!)(mod\,6) = 3(mod\,6)$.

Example: Find the remainder when 999,999 is divided by 7. We start by observing that $999,999 = 10^6 - 1$. Now, $10 = 3\,(mod\ 7)$, so $10^6 = 3^6$ $(mod\ 7)$. Since $3^2 = 9 = 2\,(mod\ 7)$, $3^6 = (3^2)^3 = 2^3 = 8 = 1\,(mod\ 7)$. Then $10^6 = 1\,(mod\ 7)$, implying that $999,999 = 0\,(mod\ 7)$, so the remainder is 0.

Example: If $a = b\ (mod\ mn)$, show that $a = b\ (mod\ m)$. Since $a = b$ $(mod\ mn)$, it follows that $mn \,|\, (a - b)$. Then $m \,|\, (a - b)$, implying that $a = b\ (mod\ m)$.

Example: Observe that $1 = 1\,(mod\,3), 2 = 2\,(mod\,3), 4 = 1\,(mod\,3), 8 = 2$ $(mod\ 3)$, $16 = 1\,(mod\ 3)$, and $32 = 2\,(mod\ 3)$. The first few powers of 2 alternate between 1 and 2 $(mod\ 3)$. Observe that $2 = -1\,(mod\ 3)$, in which case $2^n = (-1)^n\,(mod\ 3)$. Then, if n is even, we have $2^n = 1$ $(mod\ 3)$, while if n is odd, we have $2^n = -1\,(mod\ 3) = 2\,(mod\ 3)$.

Example: If a is odd, show that $a^2 = 1\,(mod\ 8)$. If a is odd, then a is either 1, 3, 5, or 7 $(mod\ 8)$. Since the squares of these numbers are 1, 9, 25, and 49, the result follows since all of these numbers are 1 $(mod\ 8)$.

Example: If p is a prime satisfying $n < p < 2n$, show that $\dbinom{2n}{n} = 0$ $(mod\,p)$. Let's begin by observing that for *any* modulus, m, we have $a = 0\,(mod\ m)$ if and only if $m \,|\, a$. Put another way, $a = 0\,(mod\ m)$ if and only if a is a multiple of m. Thus, in this example, we must show that $p \,|\, \dbinom{2n}{n}$. This becomes $p \,|\, \dfrac{(2n)!}{(n!)^2}$. Now, p is a factor of the numerator since $p < 2n$. On the other hand, p is *not* a factor of the denominator since $n < p$. It follows that $p \,|\, \dfrac{(2n)!}{(n!)^2}$, in which case $\dbinom{2n}{n} = 0(mod\,p)$.

MODULAR EQUATIONS

$ax = b$ (*mod n*), where a, b, and n are given, is a *modular* equation. An example such as $3x = 7$ (*mod* 11) should make you recall the ordinary equation $3x = 7$. We are looking for an integer solution, as 7/3 is unacceptable in modular arithmetic. You might guess that $x = 6$ is a solution, since $3 \times 6 = 18 = 7$ (*mod* 11). We shall develop a method for solving modular equations but only after we determine when they are solvable. We shall also see that there may be several *distinct* solutions, that is, solutions that are unequal *mod n*. Consider $2x = 8$ (*mod* 16). This modular equation has two distinct, or different, solutions $x = 4$ and $x = 12$ since $4 \neq 12$ (*mod* 16). We wouldn't consider $x = 20$ to be an additional solution since $20 = 4$ (*mod* 16), that is, it isn't distinct from the solution $x = 4$. The same is true for $x = 28$, 36, 44, and so on.

Here is a theorem on *linear* modular equations, that is, modular equations of first degree (as opposed to $x^2 = 1$ (*mod* 7), which is *quadratic*).

Theorem: $ax = b$ (*mod n*) is solvable if and only if $d \mid b$, where $d = gcd$ (a,n).

Proof: $ax = b$ (*mod n*) is equivalent to the statement $n \mid (ax - b)$, which is equivalent to the Diophantine equation $ax - b = kn$ in the unknowns x and k, which may be rewritten as $ax - kn = b$. As we saw in the previous chapter, this Diophantine equation is solvable if and only if $d \mid b$, where $d = gcd(a,n)$, and we are done. There will be d solutions *mod n*. Notice that if $gcd(a,n) = 1$, then $ax = b$ (*mod n*) is solvable for any b. Furthermore, the solution is unique *mod n*.   □

Example: Solve $10x = 15$ (*mod* 35). Here, $d = gcd(10,35) = 5$, and $5 \mid 15$. We thus anticipate five distinct solutions *mod* 35. To find one of them, solve $2x = 3$ (*mod* 7), yielding $x_0 = 5$. Then since $\frac{n}{d} = 7$, we have the five distinct solutions $x = 5$, 12, 19, 26, and 33. All of these numbers, by the way, satisfy $2x = 3$ (*mod* 7), but they belong to only one modular class, namely, 5 (*mod* 7), so this equation has only one *distinct* solution.

The first step in proving solvability of a modulus equation $ax = b$ (*mod n*) is computing the $gcd(a,n)$, call it d, and checking if it divides b. You saw a program to compute the gcd, but I modified it for this chapter by removing the display of messages.

Assuming that the equation is solvable, my program modifies the values a, b, and n by dividing by d. Any solution to this equation is a solution to the original equation.

I decided to write a program that systematically guesses an answer. (Systematically guessing may be considered a contradiction in terms, but I think you will agree that that is what my code does.) What does it mean for $ax = b$ (mod n)?

It means that there are integers k and x such that a times x – b is equal to k times n.

Requiring x to be an integer is important, because it is possible to solve the equation for x. The code for this is:

```
f = k*n +b;
r = f % a;
```

The next step is to check if r (the remainder after division by a) is zero. If it is, an acceptable (i.e., integral) value for x is f/a.

So, always keeping in mind that a computer can be set tedious tasks, my code tries values of k, starting with 0 and then trying 1, −1, 2, −2, 3, −3, etc. It actually tries 0 twice since my code uses a do-while loop with an internal variable k that starts off at 0 and sets kk = -k. This is not a for-loop; the incrementing is done in the body of the do { }.

The last step is to generate any other acceptable values that are distinct *mod n*. The number of such values is dependent on the value of d (remember d? it is the *gcd(a,n)* where a and n are the original a and n). If d is equal to 1, then there are not more acceptable values. Otherwise, the code generates them by adding the current value of n and doing this $(d-1)$ times.

FERMAT'S LITTLE THEOREM

Given the prime modulus p, the set of numbers $\{1, 2, 3, ..., p-1\}$ represents each nonzero congruence class of p, with no repetition. Thus, no two of the numbers in the set are congruent *mod p*. If $1 \le i < j \le p-1$ and $i = j$ *(mod p)*, we would have the ludicrous statement $p \mid (j-i)$.

We claim that the set of numbers $\{a, 2a, 3a, ..., (p-1)a\}$ also does this, that is, it represents each nonzero congruence class of p, with no repetition, provided that $gcd(a,p) = 1$, that is, as long as p does not divide *a*. To prove this, assume, to the contrary, that for some i and j satisfying $1 \le i < j \le p - 1$, we have $ia = ja \ (mod \ p)$. But when we cancel the a, this becomes $i = j \ (mod \ p)$, which is impossible.

Example: The set $\{1, 2, 3, 4\}$ represents all nonzero congruence classes *mod* 5. Now, let $a = 3$. Then the set $\{3, 6, 9, 12\}$ represents all nonzero congruence classes *mod* 5, too. The least residues are $\{3, 1, 4, 2\}$, which is a permutation of the original set $\{1, 2, 3, 4\}$.

Now, let us equate the products of the members of the two "representative" sets of the nonzero congruence classes *mod* p, since the second set is a permutation of the first, *mod* p. This yields

$$a \times 2a \times 3a \times \cdots \times (p-1)a = 1 \times 2 \times 3 \times \cdots \times (p-1)(mod \ p)$$

or

$$a^{p-1} \times (p-1)! = (p-1)!(mod \ p)$$

After canceling the common factor, $(p - 1)!$, which we have a right to do since p does not divide $(p - 1)!$, we get Fermat's little theorem:

FERMAT'S LITTLE THEOREM

Let p be prime, and let $gcd(a,p) = 1$. Then $a^{p-1} = 1 \ (mod \ p)$.   □

Example: If a is not a multiple of 5, show that $a^4 = 1 \ (mod \ 5)$. Before we prove this, note that $1^4 = 1$, $2^4 = 16$, $3^4 = 81$, and $4^4 = 256$, all of which equal 1 (*mod* 5). The proof is easy. Letting $p = 5$ in Fermat's little theorem yields $a^4 = 1 \ (mod \ 5)$ when $gcd(a,5) = 1$. Since $a = 4$ is not a multiple of 5, this condition holds.

MULTIPLICATIVE INVERSES

Given an equation such as $5x = 7$, we solve by multiplying both sides by $1/5$, since this number is the "multiplicative inverse" of 5, that is, it is the number that yields 1 when we multiply it by 5. Indeed, we get $(1/5) \times 5x = (1/5) \times 7$, which becomes $1x = (1/5) \times 7$, or $x = 7/5$. When n is prime, the next theorem tells us when we can do this with modular equations of the form $ax = b$ (*mod n*).

Theorem: Let p be prime and let a satisfy $1 \le a < p$. Then there exists a unique number b, satisfying $1 \le b < p$, such that $ab = 1$ (*mod p*), that is, a has a multiplicative inverse *mod p*.

Proof: Let's deal with the existence part first. By Fermat's theorem, $a^{p-1} = 1$ (*mod p*). Then $a \times a^{p-2} = 1$ (*mod p*). Then $b = a^{p-2}$ is a multiplicative inverse of a (*mod p*), that is, $ab = 1$ (*mod p*). As to the uniqueness part, assume that there exists another number, b', that satisfy $1 \le b' < p$, such that $ab' = 1$ (*mod p*). Then we obtain $ab - ab' = 0$ (*mod p*), implying that $a(b - b') = 0$ (*mod p*). Canceling the a yields $b - b' = 0$ (*mod p*), which implies that $b - b' = kp$. Since b and b' are between 1 and $p - 1$, this is only possible if $k = 0$, that is, if $b = b'$, thereby establishing the uniqueness (*mod p*) of the multiplicative inverse of a.   □

Note that a has infinitely many "inverses"—but only one of them is to be found in the set of "least residues" $\{1, 2, 3, \ldots, p - 1\}$. The proof of the above theorem, by the way, supplies an algorithm for finding a given number's multiplicative inverse *mod p*. The inverse of a is a^{p-2}. It follows that the *unique* solution (*mod p*) to the modular equation $ax = b$ (*mod p*) is $x = a^{p-2}b$.   □

Example: Find the multiplicative inverse of 3 (*mod 5*). Here, $a = 3$ and $p = 5$. Then the inverse is $3^3 = 27 = 2$ (*mod 5*). To check this, we observe that $3 \times 2 = 6 = 1$ (*mod 5*). Of course, we could have found the inverse of 3 by simply trying all of the candidates, namely, 2, 3, and 4. (1 is generally ruled out since it is its own inverse. In fact, $p - 1$ can even be ruled out since it is also its own inverse, since $p - 1 = -1$ (*mod p*), implying that $(p - 1)^2 = 1$ (*mod p*).)

Example: Find the multiplicative inverse of 2 (*mod 7*). Here, $a = 2$ and $p = 7$. Then the inverse is $2^5 = 32 = 4$ (*mod 7*). To check this, we observe that $2 \times 4 = 8 = 1$ (*mod 7*). Once again, it would have been easy to try each of the candidates 2, 3, 4, and 5.

WILSON'S THEOREM

The real numbers have the property that $ab = 0$ implies that $a = 0$ or $b = 0$. This permits us to cancel a *nonzero* factor c in the equation $cx = cy$.

This can be seen as follows. If $cx = cy$, then $cx - cy = 0$, which becomes $c(x - y) = 0$. Since $c \neq 0$, the other factor, $x - y$, must equal zero, that is, $x = y$. This doesn't work when $c = 0$, or we would have the following screwy "proof" that $1 = 0$. Let $a = b$. Then $a^2 = ab$. So $a^2 - b^2 = ab - b^2$, or $(a - b)(a + b) = b(a - b)$. Canceling $a - b$ yields $a + b = b$, which becomes $2b = b$, under the assumption that $a = b$. Division by b yields the ridiculous conclusion that $2 = 1$. The mistake is the cancelation of $a - b$. This number is zero.

In modular arithmetic, it is not necessarily the case that $ab = 0 \ (mod \ n)$ implies that $a = 0 \ (mod \ n)$ or $b = 0 \ (mod \ n)$. An example is afforded by $ab = 0 \ (mod \ 6)$, which is satisfied by $a = 2$ and $b = 3$. When the modulus is prime, however, this state of affairs does not persist.

Theorem: If $ab = 0 \ (mod \ p)$ where p is prime, then either $a = 0 \ (mod \ p)$ or $b = 0 \ (mod \ p)$.

Proof: $ab = 0 \ (mod \ p)$ implies that $p \mid ab$. Since p is prime, this implies that either $p \mid a$ or $p \mid b$. Then either $a = 0 \ (mod \ p)$ or $b = 0 \ (mod \ p)$, and we are done.   □

Example: If 11 does not divide a, show that $11 \mid a^5 + 1$ or $11 \mid a^5 - 1$. By Fermat's theorem, $a^{10} = 1 \ (mod \ 11)$, or $a^{10} - 1 = 0 \ (mod \ 11)$. Factoring yields $(a^5 - 1)(a^5 + 1) = 0 \ (mod \ 11)$. By the above theorem, we have $a^5 - 1 = 0 \ (mod \ 11)$ or $a^5 + 1 = 0 \ (mod \ 11)$.

Observe that $6! = 1 \times 2 \times 3 \times \cdots \times 6 \ (mod \ 7)$. We have

$$1 = 1 \ (mod \ 7)$$
$$2 \times 4 = 1 \ (mod \ 7)$$
$$3 \times 5 = 1 \ (mod \ 7)$$
$$6 = -1 \ (mod \ 7)$$

Taking the products of the left and right sides of these equations yields $6! = -1 \ (mod \ 7)$. Let's compute $10! \ (mod \ 11)$. We get

$$1 = 1 \ (mod \ 11)$$
$$2 \times 6 = 1 \ (mod \ 11)$$
$$3 \times 4 = 1 \ (mod \ 11)$$
$$5 \times 9 = 1 \ (mod \ 11)$$
$$7 \times 8 = 1 \ (mod \ 11)$$
$$10 = -1 \ (mod \ 11)$$

Taking the products of the left and right sides yields $10! = -1$ (*mod* 11). These two examples indicate a general procedure for computing $(p-1)!$ (*mod p*). Obtain a system of $(p+1)/2$ equations, the first of which is $1 = 1$ (*mod p*) and the last of which is $p - 1 = -1$ (*mod p*), while the intermediate equations are of the form $ab = 1$ (*mod p*). Each of the numbers $1, 2, \ldots, p-1$ occurs exactly once among the left members of the system of equations. (Do not fear that any of the numbers between 2 and $p - 2$ are self-inverse, that is, do not fear that for some x such that $2 \le x \le p - 2$, we have $x^2 = 1$ (*mod p*). If this were the case, we would have $x^2 - 1 = 0$ (*mod p*), or $p \mid (x-1)(x+1)$, implying that $p \mid (x-1)$, in which case $x = 1$ (*mod p*), or $p \mid (x+1)$, in which case $x = -1 = p - 1$ (*mod p*).) Taking the products of the left and right sides of the system of equations yields $(p-1)! = -1$ (*mod p*). We have just proven the following beautiful theorem.

WILSON'S THEOREM

Let p be a prime. Then $(p-1)! = -1$ (*mod p*).   □

This often appears in the equivalent formulation $p \mid (p-1)! + 1$, since $(p-1)! + 1 = 0$ (*mod p*).

Assume for a moment that Wilson's theorem is true when the modulus is not prime, and let the composite modulus n have the *proper* divisor $d \ge 2$. Then since $d \mid n$ and $n \mid (n-1)! + 1$, it follows that $d \mid (n-1)! + 1$. But since $d < n$, we have $d \mid (n-1)!$, which is impossible. So Wilson's theorem only applies to primes. It may, therefore, be restated as follows.

WILSON'S THEOREM (2ND VERSION)

Let $n \ge 2$. Then $(n-1)! = -1$ (*mod n*) if and only if n is prime.   □

When $n = 4$, note that $3! = 6 = 2$ (*mod* 4). On the other hand, when $n > 4$ and n is composite, it can be shown that $(n-1)! = 0$ (*mod n*). For example, $5! = 120 = 0$ (*mod* 6).

Example: You have no calculator and you need the remainder when $100!$ is divided by 101. Using Wilson's theorem, you obtain $100! = -1$ (*mod* 101), from which it follows that $100! = 100$ (*mod* 101). Then the remainder is 100.

Example: Now, you seek the remainder when 99! is divided by 101. Since $100! = 99! \times 100$, you obtain $99! \times 100 = -1$ (*mod* 101). You then realize that $100 = -1$ (*mod* 101) and substitute this into the previous equation, getting $99! \times (-1) = -1$ (*mod* 101), which, after canceling, becomes $99! = 1$ (*mod* 101). So the remainder is 1.

SQUARES AND QUADRATIC RESIDUES

Consider the squares of the four nonzero least residues, 1, 2, 3, and 4, of the modulus 5. They are 1, 4, 9, and 16, respectively. When we reduce these numbers, *mod* 5 (i.e., we compute their least residues), we get 1, 4, 4, and 1, respectively.

Let's consider another (odd) prime modulus, say, 7. The six nonzero least residues, 1, 2, 3, 4, 5, and 6, when squared and reduced *mod* 7 yield 1, 4, 2, 2, 4, and 1, respectively.

Let's consider just one more (odd) prime modulus, 11. The 10 nonzero least residues, 1 through 10, when squared and reduced *mod* 11 yield 1, 4, 9, 5, 3, 3, 5, 9, 4, and 1, respectively.

Notice that in each of the above odd primes, each reduced square occurs twice. As a consequence, only half of the nonzero least residues have "square roots." (These least residues are called *quadratic residues*.) Put another way, when $p = 5$, 7, or 11, the modular equation $x^2 = a$ (*mod p*), which calls for the square root of a, has a solution for only half of the possible values of a. Thus, for example, $x^2 = a$ (*mod* 5) has a solution when a is 1 or 4, but has no solution when $a = 2$ or 3, that is, 1 and 4 are quadratic residues, while 2 and 3 are not. Moreover, the nonzero a's for which the modular equation $x^2 = a$ (*mod p*) has a solution are quite special. There are two solutions for each such a, while there is no solution for the other a's.

The three lists of squares we obtained are palindromes. A *palindrome* is a word or phrase that doesn't change when read backward. An example is "Eva can I stab bats in a cave." The list of squares *mod* 11 is 1, 4, 9, 5, 3, 3, 5, 9, 4, 1, which is certainly a palindrome. If you square the nonzero residues, *mod* 11, in ascending order, that is, 1, 2, 3, …, 10 and your friend squares them in descending order, that is, 10, 9, 8, …, 1, both of you will obtain analogous results.

Are these observations for the odd primes 5, 7, and 11 coincidental? Hardly. When p is an odd prime, there are an even number of nonzero residues, 1, 2, …, $p - 2$, and $p - 1$. Now, for each a, satisfying $1 \le a \le p - 1$, we find that $p - a = -a$ (*mod p*). Squaring both sides yields $(p - a)^2 = a^2$ (*mod p*). For example, when $p = 11$ and $a = 3$, we find that $8^2 = 3^2$ (*mod* 11).

You might wonder whether more than two nonzero least residues of an odd prime can have the same square. This is impossible. Suppose the three nonzero least residues, a, b, and c, of the odd prime p satisfy the equations $a^2 = b^2$ (*mod p*) and $a^2 = c^2$ (*mod p*). Then the first equation becomes $a^2 - b^2 = 0$ (*mod p*), or $(a-b)(a+b) = 0$ (*mod p*), implying by a previous theorem that either $b = a$ (*mod p*) or $b = -a$ (*mod p*). By the same reasoning, the equation $a^2 = c^2$ (*mod p*) implies that either $c = a$ (*mod p*) or $c = -a$ (*mod p*). Then the only solutions are a and $-a$.

On the other hand, the above analysis is not true *mod n* where n is composite. We have, for example, $1^2 = 3^2 = 5^2 = 7^2 = 1$ (*mod 8*). Here, the four different inputs 1, 3, 5, and 7 have the same output, namely, 1 (*mod 8*).

In the set of real numbers, -1 has no square root. When is -1 a quadratic residue *mod p*? In other words, for which odd primes does the equation $x^2 = -1$ (*mod p*) have a solution? Recall that odd primes have one of the forms $4k + 1$ or $4k + 3$. Alternatively, any odd prime satisfies exactly one of the equations $p = 1$ (*mod 4*) or $p = 3$ (*mod 4*). Let's assume that $a^2 = -1$ (*mod p*). Now, $gcd(a,p) = 1$, in which case we may invoke Fermat's little theorem. We get the following equation:

$$1 = a^{p-1} = \left(a^2\right)^{\frac{p-1}{2}} = (-1)^{\frac{p-1}{2}} (mod\ p)$$

or

$$1 = (-1)^{\frac{p-1}{2}}$$

This implies that $\dfrac{p-1}{2}$ is even, since -1 raised to an odd power is -1. Then $\dfrac{p-1}{2} = 2k$, for some k, implying that $p - 1 = 4k$ and $p = 4k + 1$. We infer that -1 is not a quadratic residue when p is a prime of the form $4k + 3$. Thus, for example, $x^2 = -1$ (*mod 7*) has no solution. Now, if p is of the form $4k + 1$, is -1 guaranteed to be a quadratic residue? We have the following reasoning. Observe that

$$(p-1)! = 1 \times 2 \times 3 \times \cdots \times \frac{p-1}{2}$$
$$\times \left(p - \frac{p-1}{2}\right) \times \cdots \times (p-3) \times (p-2) \times (p-1)$$

Since $p - a = -a$ (*mod p*), we can rewrite this as

$$(p-1)! = 1 \times (-1) \times 2 \times (-2) \times 3 \times (-3) \times \cdots$$
$$\times \frac{p-1}{2} \times \left(-\frac{p-1}{2}\right) (mod\ p)$$

Notice that there are $\dfrac{p-1}{2}$ negative signs that permits us to rewrite this, once again, as

$$(p-1)! = \left[\left(\frac{p-1}{2}\right)!\right]^2 (-1)^{\frac{p-1}{2}} (mod\ p) \qquad (5.1)$$

Since we are assuming that $p = 4k + 1$, it follows that $\dfrac{p-1}{2} = 2k$ and we can dispense with the last factor on the right side of (5.1). Furthermore, by Wilson's theorem, $(p-1)! = -1$ $(mod\ p)$. Then (5.1) becomes

$$-1 = [(2k)!]^2 (mod\ p)$$

which shows that -1 is a quadratic residue when $p = 4k + 1$. When $p = 13$, for example, we have $k = 3$, and $(2k)! = 6! = 720 = 5$ $(mod\ 13)$. Then $5^2 = 25 = -1$ $(mod\ 13)$, establishing that -1 is indeed a quadratic residue, $mod\ 13$. To summarize this discussion, we have the following theorem.

Theorem: Let p be an odd prime. Then the equation $x^2 = -1$ $(mod\ p)$ has a solution if and only if $p = 1$ $(mod\ 4)$.   □

LAGRANGE'S THEOREM

Recall from Chapter 4 that given integers a and b, where a is positive, there exists a pair of numbers, q and r, where $0 \le r < a$, such that $b = qa + r$. The number q is the *quotient* and r is the *remainder* when we divide b by a. An analogous equation applies to polynomials with integer coefficients. When dividing $f(x)$ by $g(x)$, we obtain $f(x) = q(x)g(x) + r(x)$, where $q(x)$ is the *quotient* and $r(x)$ is the *remainder*. The degree of $r(x)$ is less than the degree of $g(x)$. Furthermore, $q(x)$ and $r(x)$ have integer coefficients. Let's look at an example to make this clear.

Example: The division of $f(x) = 2x^3 + 3x + 2$ by $g(x) = x^2 + 1$ yields the quotient $2x$ and the remainder $x + 2$. The division can be written $2x^3 + 3x + 1 = (2x)(x^2 + 1) + (x + 2)$. Note that the degree of the remainder is 1, which is less than the degree, 2, of $g(x)$.

When $g(x) = x - a$, the division equation becomes $f(x) = q(x)(x - a) + r$, where r is an integer. Letting $x = a$ in this equation shows us that $r = f(a)$. If a is a *root* of $f(x)$, that is, if $f(a) = 0$, then we have

$$f(x) = q(x)(x - a)$$

implying that the degree of $q(x)$ is $n - 1$, where n is the degree of $f(x)$.

We state Lagrange's theorem on the number of solutions of modular equations involving polynomials when the modulus is prime.

LAGRANGE'S THEOREM

Let p be prime and let $f(x) = a_n x^n + a_{n-1} x^{n-1} + \cdots + a_1 x + a_0$, where $a_n \neq 0$ (*mod p*); all of the coefficients are integers, and $n \geq 1$. Then the equation $f(x) = 0$ (*mod p*) has at most n distinct solutions *mod p*. □

Corollary: Let p be prime. Then $x^2 = 0$ (*mod p*) has at most two distinct solutions *mod p*. □

Example: Find all solution to the equation $x^3 = 1$ (*mod* 13). Obviously, one solution is $x = 1$. Now, $3^3 = 27 = 1$ (*mod* 13), so $x = 3$ is a second solution. Furthermore, $9^3 = (3^2)^3 = (3^3)^2 = 1^2 = 1$ (*mod* 13), so $x = 9$ is a third solution. By Lagrange's theorem, there can be no more than three solutions.

Corollary: Let p be prime and let $d \mid p - 1$. Then the equation $x^d - 1 = 0$ (*mod p*) has exactly d distinct solutions *mod p*.

Proof: $d \mid p - 1$ implies that $p - 1 = dk$. The equation $x^{p-1} - 1 = 0$ (*mod p*) has exactly $p - 1$ solutions *mod p*, namely, $x = 1$, $x = 2$, ..., and $x = p - 1$. This equation can be written $x^{dk} - 1 = 0$ (*mod p*). Letting $y = x^d$, we have $x^{dk} - 1 = y^k - 1 = (y - 1)(y^{k-1} + y^{k-2} + \cdots + y + 1) = (x^d - 1)(x^{d(k-1)} + x^{d(k-2)} + \cdots + x^d + 1)$. Then $x^{p-1} - 1 = 0$ (*mod p*) can be rewritten as $(x^d - 1)(x^{d(k-1)} + x^{d(k-2)} + \cdots + x^d + 1) = 0$. The second factor is a polynomial of degree $d(k - 1)$ and can have at most $d(k - 1)$ roots (*mod p*). Then the first factor, $x^d - 1$, must have at least $dk - [d(k - 1)]$, or d roots. By Lagrange's theorem, the equation $x^d - 1 = 0$ (*mod p*) has at most d roots, and the proof of the corollary is complete. □

Example: Show, without finding them, that the equation $x^4 = 1$ (*mod* 13) has exactly four solutions. Then find the solutions. By the above corollary, since $4 \mid (13 - 1)$, the equation has exactly four solutions. Obviously 1 and 12 are solutions. Now, since $5^2 = 25 = -1 \ (mod \ 13)$, we have $5^4 = (5^2)^2 = 1 \ (mod \ 13)$. So 5 is another solution. Finally, we have $8^4 = (13 - 5)^4 = 5^4 = 1 \ (mod \ 13)$, and the fourth solution is 8.

REDUCED PYTHAGOREAN TRIPLES

The Pythagorean theorem was first proven in the sixth century B.C. by Pythagoras, though it was known to the Mesopotamian mathematicians of 2000 b.c. While the theorem belongs to geometry, it is of interest to number theorists when a, b, and c are positive integers that satisfy $a^2 + b^2 = c^2$. It is remarkable that the sum of two squares can be a square. A set of positive integers $\{a, b, c\}$ satisfying the equation $a^2 + b^2 = c^2$ is called a *Pythagorean triple*.

Are there infinitely many Pythagorean triples? Yes. For any k, we have $\{3k, 4k, 5k\}$. After all, $(3k)^2 + (4k)^2 = 9k^2 + 16k^2 = 25k^2 = (5k)^2$. We are disappointed. All right triangles with sides of lengths $3k$, $4k$, and $5k$ are similar to the original "3, 4, 5" right triangle. A *reduced Pythagorean triple*, or *RPT*, consists of a Pythagorean triple, $\{a, b, c\}$, such that a, b, and c have no common factor. Each *RPT* adds something new to our list of such triples and produces a uniquely shaped right triangle. This raises two questions. (i) Are there infinitely many *RPT*s? (ii) How do we find them? We require several small theorems, called *lemmas*, before we get to the answers.

Lemma 1: Let $\{a, b, c\}$ be an *RPT*. Then a and b have opposite parity.

Proof: If a and b are even, then $a = 2k$ and $b = 2j$. Then $c^2 = 4k^2 + 4j^2$ in which case $4 \mid c^2$. This implies that c^2 is even. Then c is even, contradicting the assumption that $\{a, b, c\}$ is an *RPT*. On the other hand, if a and b are odd, then $a = 2k + 1$ and $b = 2j + 1$. Then $c^2 = (2k + 1)^2 + (2j + 1)^2 = 4k^2 + 4k + 1 + 4j^2 + 4j + 1 = 2 \ (mod \ 4)$. This is impossible, since we proved earlier in this chapter that for any nonzero s, we have $s^2 = 0$ or $1 \ (mod \ 4)$. $\square$

Lemma 2: Let $gcd(r,s) = 1$ and suppose that $rs = t^2$. Then r and s are squares.

Proof: Since $gcd(r,s) = 1$, it follows that the prime decompositions of r and s have no common primes. On the other hand, the exponents of the primes in the prime decomposition of t^2 are even. Then the same goes for the exponents of the primes of r and s, in which case they are squares. □

Let $\{a, b, c\}$ be an *RPT*. By Lemma 1, let a be even and let b be odd. Then c must be odd, since c^2 is the sum of the even number a^2 and the odd number b^2. Then $c - b$ and $c + b$ are even. Then let $c - b = 2u$ and $c + b = 2v$. We then have $a^2 = c^2 - b^2 = (c - b)(c + b) = 4uv$, which implies that $(a/2)^2 = uv$. Claim: u and v are relatively prime. To see this, note that the equations $c - b = 2u$ and $c + b = 2v$ imply that $c = v + u$ and $b = v - u$. If u and v had a common divisor, so would b and c. Then a would also have this divisor, contradicting the fact that $\{a, b, c\}$ is an *RPT*. Since $(a/2)^2 = uv$ and $\gcd(u,v) = 1$, it follows by Lemma 2 that u and v are squares. Then let $u = t^2$ and let $v = s^2$. Then $a^2 = 4uv = 4s^2t^2$, implying that $a = 2st$. Furthermore, $\gcd(s,t) = 1$. The equations $c = v + u$ and $b = v - u$ become $c = s^2 + t^2$ and $b = s^2 - t^2$. Finally, note that the equation $b = s^2 - t^2$ implies that s and t have opposite parity or b would be even. To summarize this discussion, the following theorem outlines a procedure that will generate infinitely many *RPT*s.

Theorem: Select a pair of numbers, s and t, such that (i) $\gcd(s,t) = 1$, (ii) $s > t$, and (iii) s and t have opposite parity. Then

$$a = 2st$$
$$b = s^2 - t^2$$
$$c = s^2 + t^2$$

generates an *RPT* and all *RPT*s are generated this way.

Proof: It is easy to see that $a^2 + b^2 = (2st)^2 + (s^2 - t^2)^2 = s^4 + 2s^2t^2 + t^4 = (s^2 + t^2)^2 = c^2$. Furthermore, since $\gcd(s,t) = 1$, $\{a, b, c\}$ is an *RPT*. The discussion prior to the theorem establishes the claim that all *RPT*s are generated this way. □

When $s = 2$ and $t = 1$, we obtain $a = 4, b = 3$, and $c = 5$, while when $s = 3$ and $t = 2$, we obtain $a = 12$, $b = 5$, and $c = 13$. On the other hand, when $s = 3$ and $t = 1$, we obtain $a = 6$, $b = 8$, and $c = 10$, a multiple of the *RPT*, $\{3, 4, 5\}$.

Recall from Chapter 2 that if $x = a^2 + b^2$ and $y = c^2 + d^2$, then $xy = (a^2 + b^2)(c^2 + d^2) = (ac - bd)^2 + (ad + bc)^2$. If $a = c$ and $b = d$, this becomes $(a^2 + b^2)^2 = (a^2 - b^2)^2 + (2ab)^2$.

Theorem: There are infinitely many *RPT*s for which $c - a = 1$.

Proof: This requires that $s^2 + t^2 - 2st = 1$, or $(s - t)^2 = 1$. Then let $s = t + 1$. □

It is harder to produce infinitely many *RPT*s for which $a - b = 1$, the possibility of which is guaranteed by our next theorem.

Theorem: There are infinitely many *RPT*s for which $a - b = 1$.

Proof: This requires that $2st - (s^2 - t^2) = 1$, or $2st - s^2 + t^2 = 1$. This becomes $s^2 + 2st + t^2 - 2s^2 = 1$, or finally, $(s + t)^2 - 2s^2 = 1$. Letting $x = s + t$ and $y = s$, we have the Pell equation $x^2 - 2y^2 = 1$, which was solved in Chapter 2. It was shown there that this equation has infinitely many solutions. The solution $x = 17$, $y = 12$, for example, yields $s = 12$ and $t = 5$, which in turn yields the *RPT* $a = 120$, $b = 119$, and $c = 169$. □

Here is a theorem whose proof uses *RPT*s. First, we require a definition. A sequence of k numbers, $a_1, a_2, \ldots, a_k$, is an *arithmetic progression* if there exists a number d, the *difference*, such that $a_2 - a_1 = a_3 - a_2 = \ldots = a_k - a_{k-1} = d$. The number of terms, k, is the *length* of the progression. The sequence 3, 7, 11 is an arithmetic progression of length 3 with a difference of 4.

Theorem: There exist infinitely many arithmetic progressions of length three whose terms are squares.

Proof: Call the squares x^2, y^2, and z^2. Then $y^2 - x^2 = z^2 - y^2$ so $2y^2 = x^2 + z^2$. Now, let $x = u - v$ and let $z = u + v$. Then $2y^2 = x^2 + z^2 = (u - v)^2 + (u + v)^2 = 2(u^2 + v^2)$, implying that $y^2 = u^2 + v^2$. This states that u, v, and y form a Pythagorean triple. Then let $u = 2st$, $v = s^2 - t^2$, and $y = s^2 + t^2$. So we have infinitely many solutions:

$$x = 2st - s^2 + t^2$$
$$y = s^2 + t^2$$
$$z = 2st + s^2 - t^2 \qquad \qquad □$$

When $s = 2$ and $t = 1$, for example, we get $x = 1$, $y = 5$, and $z = 7$, yielding the arithmetic progression 1, 25, 49, with common difference 24.

CHINESE REMAINDER THEOREM

Here is an interesting problem asked by a Chinese mathematician 2000 years ago. Find a number whose remainders when divided by 3, 5, and 7, are 2, 3, and 2, respectively. In other words, solve the simultaneous modular equations:

$$x = 2 \pmod 3$$
$$x = 3 \pmod 5$$
$$x = 2 \pmod 7$$

A solution is $x = 23$, as we can quickly verify. But how do we obtain a solution? Is the solution unique, or are there more solutions? Are there infinitely many solutions? Before we present the so-called Chinese remainder theorem, we require a definition. The integers $a_1, a_2, \ldots, a_k$ are called *pairwise relatively prime* if $gcd(a_i, a_j) = 1$ for $i \neq j$. In other words, any two of these numbers are relatively prime. This is stronger than the following statement: the integers $a_1, a_2, \ldots, a_k$ have no common factor greater than 1. Thus while the integers 3, 6, 7, and 8 have no common factor greater than 1, they are *not* pairwise relatively prime, since $gcd(3,6) = 3$. On the other hand, the integers 2, 5, 9, and 77 are pairwise relatively prime. (It follows that they also obey the weaker condition, that is, they have no common factor greater than 1.) Here is the theorem.

CHINESE REMAINDER THEOREM

Let the positive integers $n_1, n_2, \ldots, n_k$ be pairwise relatively prime, and let $n = n_1 \times n_2 \times \cdots \times n_k$. Then the system of modular equations

$$x = a_1 \pmod{n_1}$$
$$x = a_2 \pmod{n_2}$$
$$\vdots$$
$$x = a_k \pmod{n_k}$$

has a simultaneous solution that is unique *mod n*.

Proof: For each j between 1 and k, let $m_j = \frac{n}{n_j}$. In other words, m_j is the product of the n_i's except for n_j. For example, $m_1 = n_2 \times n_3 \times \cdots \times n_k$ and $m_2 = n_1 \times n_3 \times \cdots \times n_k$. Note that $gcd(m_j, n_j) = 1$, since m_j is formed by deleting n_j from the product of the n_i's, and n_j is relatively prime to each of the "surviving" n_i's. Consider the equation $m_j x = 1 \pmod{n_j}$. By a corollary stated earlier in this chapter, it has a solution, which we shall denote by x_j. Consider the number $X = a_1 m_1 x_1 + a_2 m_2 x_2 + \cdots + a_k m_k x_k$. To see that this is a simultaneous solution, note that $m_j = 0 \pmod{n_i}$

when $i \neq j$, since in this case $n_i \mid m_j$. Then for each j between 1 and k, we have $X = a_j m_j x_j \ (mod\ n_j)$. But $m_j x_j = 1 \ (mod\ n_j)$, in which case we have for each j between 1 and k, $X = a_j \ (mod\ n_j)$. So X is a simultaneous solution. Now, if Y is any other simultaneous solution, we have $X = Y \ (mod\ n_j)$ for $j = 1, 2, ..., k$. Then $n_j \mid X - Y$ for $j = 1, 2, ..., k$. Since the n_j's are pairwise relatively prime, it follows by a theorem in the previous chapter that $n \mid (X - Y)$, that is, that $X = Y \ (mod\ n)$. Then there is a unique solution $mod\ n$. $\qquad\qquad\qquad\qquad\qquad\qquad\qquad\qquad\qquad\qquad\qquad\qquad\qquad$ □

As a consequence of the last paragraph, there are infinitely many solutions, that is, all members of the congruence class of $X \ mod\ n$.

Example: Solve the system

$$x = 2(mod\ 3)$$
$$x = 3(mod\ 5)$$
$$x = 2(mod\ 7)$$

3, 5, and 7 are pairwise relatively prime, so there exists a solution (which is unique $mod\ n$, where $n = 3 \times 5 \times 7 = 105$). Now, $m_1 = 105/3 = 35$, $m_2 = 105/5 = 21$, and $m_3 = 105/7 = 15$. Then we must solve the modular equations

$$35x = 1(mod\ 3)$$
$$21x = 1(mod\ 5)$$
$$15x = 1(mod\ 7)$$

yielding $x_1 = 2$, $x_2 = 1$, and $x_3 = 1$. Then $X = 2 \times 35 \times 2 + 3 \times 21 \times 1 + 2 \times 15 \times 1 = 233$. This is 23 $(mod\ 105)$.

EXERCISES

An asterisk (∗) indicates that the exercise can be developed into a research project.

1   Show that $n^3 = 0$ or $\pm 1 \ (mod\ 9)$. The first few cubes 1, 8, 27, 64, 125, and 216 suggest the pattern 0, 1, −1, 0, 1, −1, ... Does this pattern persist for the first 100 cubes?

2   Show that $n^3 = 0$ or $\pm 1 \ (mod\ 7)$ for $n \leq 100$.

3 Show that $n^4 = 0$ or 1 (*mod* 5) for $n \le 100$. Then prove it.

4 *Show that $x^4 + y^4 + z^4$ is not divisible by 5 unless $x = y = z = 0$ (*mod* 5).

5 Solve $5x = 2$ (*mod* 26), $7x = 3$ (*mod* 13), $4x = 14$ (*mod* 27), and $2x = 5$ (*mod* 7).

6 Find all seven distinct solutions of $14x = 21$ (*mod* 35).

7 Find all four distinct solutions of $8x = 12$ (*mod* 36).

8 Prove properties 1 through 7 listed in the section titled "Congruence Classes Mod k."

9 Find the multiplicative inverses of the numbers from 1 through 10 (*mod* 11) and use these inverses to solve $3x = 2$ (*mod* 11), $5x = 7$ (*mod* 11), and $6x = 9$ (*mod* 11).

10 *The numbers 11, 111, 1111, … are called *repunits*. Show that no repunit is a square by showing that if n is a repunit, then $n = 3$ (*mod* 4).

11 Show that $r^2 = s^2$ (*mod* n), where $1 \le r \le s < n$ does not imply that $r = s$ (*mod* n).

12 Solve the equation $x^5 = 1$ (*mod* 11). It has exactly five solutions.

13 Find 10 *RPT*s for which $b - a = 1$.

14 *For a and b less than 20, show that exactly one of a, b, $a + b$, and $a - b$ in any *RPT* is divisible by 7. Does this pattern persist?

15 Show that the numerator and denominator of the sum of the reciprocals of consecutive odd numbers are the legs of an *RPT*. For example, $\frac{1}{3} + \frac{1}{5} = \frac{8}{15}$, and 8 and 15 belong to the *RPT* (8, 15, 17).

16 *Find 10 arithmetic progressions of length three whose terms are squares and whose first term is 1. An example is 1, 841, 1681. (This is 1^2, 29^2, 41^2.)

17 Show that $S = 1 + \frac{1}{2} + \frac{1}{3} + \cdots + \frac{1}{n}$ is not an integer for $2 \le n \le 20$. In fact, it is never an integer after 1.

18 What is the remainder when 2^{100} is divided by 7? Hint: $2^3 = 1$ (*mod* 7).

19 Use number theory to find the remainder when $\sum_{k=1}^{1000} k!$ is divided by 24. How much running time is involved if this is done by computing the sum and then dividing by 24?

20 *Show that $1^{p-1} + 2^{p-1} + 3^{p-1} + \cdots + (p-1)^{p-1} = -1 (mod\ p)$, where p is a prime less than 50.

21 Show that none of the first 1000 triangular number ends with the digits 2, 4, 7, or 9.

22 Is t_n ever the sum of three consecutive squares for $n \le 20$?

23 *Show that no x simultaneously satisfies $x = 5\ (mod\ 6)$ and $x = 7\ (mod\ 15)$ for $x \le 1000$.

24 Solve the simultaneous equations $x = 1\ (mod\ 3)$, $x = 1\ (mod\ 5)$, and $x = 1\ (mod\ 7)$.

25 Using the binomial theorem for $(1 + 1)^p$, show that $2^p = 2\ mod\ p$, where p is a prime.

26 Write a program to determine the truth or falsehood of the modular equation $a = b\ (mod\ n)$, given a, b, and n. *See the program at the end of the chapter.*

27 By squaring the numbers from 1 to 100, find the last two digits (i.e., the unit's and ten's digits) of any square. This information will be useful in showing that a given number is not a square. No square ends with 15, for example, showing that 67,915 is not a square.

28 Write a program that determines whether $ax = b\ (mod\ n)$ is solvable. If it is solvable, your program should yield a complete set of distinct solutions, each of which is a least residue $mod\ n$. See the program at the end of the chapter. *You are encouraged to try an alternate approach and compare.*

29 Find numbers a and p such that p is prime, a and p are relatively prime, and $a^{p-1} = 1\ (mod\ p^2)$. As an example, when $a = 3$ and $p = 11$, we have $3^{10} = 1\ (mod\ 121)$. Note that it is usually not the case that $a^{p-1} = 1\ (mod\ p^2)$. Fermat's little theorem only guarantees that $a^{p-1} = 1\ (mod\ p)$. Use the following values of p: 3, 5, 7, 11, and 13.

30 *While Wilson's theorem guarantees that $p \mid (p-1)! + 1$, sometimes, $p^2 \mid (p-1)! + 1$. For which primes $p \le 17$ is this true?

31 Note that $(n-1)! + 1$ is a square when $n = 5, 6$, and 8. Verify that there are no other solutions for $n \le 10$.

32 Write a program that produces all *RPT*s for which $a - b = 1$ and $a < N$, where N is a given positive integer.

Testing Truth or Falseness of a Modulo Equation

```
<!DOCTYPE HTML>
<html>
<meta charset="UTF-8">
<head>
<title>Testing mod eq</title>
<script>
var placeref;

//ax = b (mod n) is solvable if and only
//if d | b, where d = gcd(a,n).
function truefalse(){
 placeref = document.getElementById("place");
 placeref.innerHTML = "";
 a = parseInt(document.f.numA.value);
 n = parseInt(document.f.numn.value);
 b = parseInt(document.f.numB.value);
 ans = (a-b) % n;
 if (ans==0){
     placeref.innerHTML = "The equation is true."
 }
 else {
     placeref.innerHTML = "The equation is false."
 }
 return false;
}
</script>
</head>
<body>
Enter 3 integers for mod equation a = b mod(n) <br/>
<form name="f" onsubmit="return truefalse();">
 a: <input type="number" name="numA" value=""/>
 
 b: <input type="number" name="numB" value=""/>
 
```

```
 n: <input type="number" name="numn" value=""/><br/>
  <input type="submit" value="Enter"/>
</form>
<p id="place">
  Messages will go here.
</p>
</body>
</html>
```

Solvability of Modulo Equation and Finding Solutions

```
<!DOCTYPE HTML>
<html>
<meta charset="UTF-8">
<head>
<title>Solving mod eq</title>
<script>
 var placeref;

 //ax = b (mod n) is solvable if and only if d | b, where
//d = gcd(a,n).
function solvable(){
 placeref = document.getElementById("place");
 placeref.innerHTML = "";
 a = parseInt(document.f.numA.value);
 n = parseInt(document.f.numn.value);
 d = gcd(a,n);
 b = parseInt(document.f.numB.value);
 if ((b % d)==0) {
    placeref.innerHTML="Solvable";
    //now to solve it
    a = a / d;
    b = b / d;
    n = n / d;
    //determine a solution
    messages = "Solving "+String(a)+"x = "+String(b)+
       " (mod "+String(n)+") <br/>";
    placeref.innerHTML = messages;
    k=-1;
    do {
      k++;
      f = (k*n+b);
      r = f % a;
```

```
    // also trying -k. Note when k is zero, this is
//redundant
      kk = -k;
      ff = (kk*n+b);
      rr = ff % a;
    }
    while ((r!=0) && (rr!=0));
    if (r==0) {
      x = f / a;
    }
    else {
      x = ff / a;
    }
    if (d==1) {
        messages+="<br/> Unique solution (mod "+
            String(n)+") is " +String(x);
        placeref.innerHTML = messages;
    }
    else {
    messages+= "<br/> Initial solution x0 is
"+String(x);
    placeref.innerHTML = messages;

    for (j=1;j<d;j++){
      x = x+n;
      messages+="<br/> Another solution is
"+String(x);
      placeref.innerHTML = messages;
     }
    }
    return false;
 }
 else {
    placeref.innerHTML="not solvable";

 }
 return false;
}
function gcd(a,b){
  var holder;
  var r;
  if (b < a) {
```

```
    //switch

    holder = a;
    a = b;
    b = holder;
    }
    r = b % a;
    if (r==0){

        return a;
    }

    else {
        return gcd(r,a);

    }
}


</script>
</head>
<body>
Enter 3 integers for mod equation ax = b mod(n) <br/>
<form name="f" onsubmit="return solvable();">
 a: <input type="number" name="numA" value=""/>
 
 b: <input type="number" name="numB" value=""/>
 
 n: <input type="number" name="numn" value=""/><br/>
  <input type="submit" value="Enter"/>
</form>
<p id="place">
  Messages will go here.
</p>
</body>
</html>
```

6

NUMBER THEORETIC FUNCTIONS

We examine a class of interesting functions used in number theory.

THE *TAU* FUNCTION

The function $\tau(n)$ counts how many divisors n has. This count includes 1 and n. (τ is a Greek letter and is called "tau.") The first few values are $\tau(1) = 1$, $\tau(2) = 2$, $\tau(3) = 2$, $\tau(4) = 3$, $\tau(5) = 2$, $\tau(6) = 4$, $\tau(7) = 2$, $\tau(8) = 4$, $\tau(9) = 3$, and $\tau(10) = 4$.

Our first tau function is based on the following definition: counting the factors, starting at 1 and ending with the number itself. (Repeat warning: the decision in counting factors on including one or including the number itself is dependent on the situation and can be tricky to check.) The program consists of two functions. The one named `tau` performs the interactions (input/output) with the user. It invokes `produceFactors` that returns an array. The value of tau is the length of this array.

Elementary Number Theory with Programming, First Edition. Marty Lewinter and Jeanine Meyer.
© 2016 John Wiley & Sons, Inc. Published 2016 by John Wiley & Sons, Inc.

Let n have the prime decomposition $p_1^{e_1}p_2^{e_2}...p_r^{e_r}$. It is easy to see that m is a divisor of n if and only if $m = p_1^{c_1}p_2^{c_2}...p_r^{c_r}$ where $0 \le c_1 \le e_1, 0 \le c_2 \le e_2, 0 \le c_3 \le e_3, ..., 0 \le c_r \le e_r$. (This can be stated more tersely as "$0 \le c_i \le e_i$ for $i = 1, 2, ..., r$.") As an example, the divisors of $2^3 3^2 7^1 = 8 \times 9 \times 7 = 504$ have the form $2^a 3^b 7^c$, where $a = 0, 1, 2,$ or 3; $b = 0, 1,$ or 2; and $c = 0$ or 1. In general, there are $e_1 + 1$ values for c_1 (including 0); $e_2 + 1$ values for c_2, ...; and $e_r + 1$ values for c_r. This implies the following simple formula for computing $\tau(n)$:

$$\tau(n) = (e_1 + 1)(e_2 + 1)(e_3 + 1)...(e_r + 1) \qquad (6.1)$$

Example: Since $20 = 2^2 5^1$, $\tau(20) = (2 + 1)(1 + 1) = 3 \times 2 = 6$.

Example: Since $840 = 2^3 3^1 5^1 7^1$, $\tau(840) = (3 + 1)(1 + 1)(1 + 1)(1 + 1) = 4 \times 2 \times 2 \times 2 = 32$. This is the largest value of τ for all integers less than 1000.

The second tau program is based on using the formula from prime decomposition and provides an opportunity to tell you (perhaps, remind you) of how to bring other code into a JavaScript program. All programming languages have features for doing this, though the exact mechanisms vary. In JavaScript, we use an external script element:

```
<script type="text/javascript"
src="primeDecomposition.js">
</script>
```

The file *primeDecomposition.js* contains code much like the earlier (see Chapter 4) program. One difference is that the program `build-PrimeDecomposition` returns an array. This is an array of arrays with the second element (index equal to 1) of each array being the exponent of the prime. The function `tauFF` (in the file *tauFromFormula.html*) invokes the function `buildPrimeDecomposition` and uses the result to calculate the formula:

```
table = buildPrimeDecomposition(n);
  t = 1;
  for (var i=0;i<table.length;i++){
```

```
    t = t * (1+table[i][1]);
}
```

This explanation is longer than the *tauFromFormula.html* file. Look at the end of the chapter for the code for both files.

Example: $\tau(n) = 2$ if and only if n is prime. If $n = p$, where p is a prime, it is obvious that $\tau(n) = \tau(p^1) = 2$. Moreover, if $\tau(p_1^{e_1} p_2^{e_2} \ldots p_r^{e_r}) = (e_1 + 1)(e_2 + 1)(e_3 + 1) \ldots (e_r + 1) = 2$, then $e_j = 1$, for some specific j such that $1 \leq j \leq r$, while $e_i = 0$ for $i \neq j$.

Example: Find the smallest number n for which $\tau(n) = 4$. Let us first determine the possible prime decompositions for n. We have, from (6.1), $4 = (e_1 + 1)(e_2 + 1)(e_3 + 1) \ldots (e_r + 1)$. Since $4 = 1 \times 4 = 2 \times 2$, we have the solutions $n = p^3$ and $n = pq$, where p and q are distinct primes. If $p = 2$, $n = 8$, while if $p = 2$ and $q = 3$, $n = 6$. Clearly, 6 is the smallest number for which $\tau(n) = 4$.

Here is an interesting theorem about τ.

Theorem: $\tau(n)$ is odd if and only if n is a square.

Proof: Let $n = s^2$. Then for each divisor d such that $1 \leq d < s$, there is a unique divisor $d' = \dfrac{n}{d}$, such that $s < d' \leq n$. (For example, if $s = 10$ and $n = 10^2 = 100$, then given the divisor $4 < 10$, there exists the unique divisor $\dfrac{100}{4} = 25 > 10$. Of course, $4 \times 25 = 100$.) Furthermore, for each divisor h such that $s < h < n$, there is a unique divisor $h' = \dfrac{n}{h}$, such that $1 \leq h' < s$. In other words, the divisors of n, with the exception of s, occur in pairs—one less than s and the other greater than s. Adding the divisor s to the list yields an odd number of divisors.

On the other hand, if n is not a square, the above argument indicates that the total number of divisors is *even* since the number of divisors greater than $\sqrt{n}$ equals the number of divisors less than $\sqrt{n}$. (This follows even more simply by observing that if $ab = n$ and $a < b$, then $a < \sqrt{n}$ and $b > \sqrt{n}$.) □

Let us compute the product of the divisors of n. If n is not a square, there are an even number of divisors, half of which are less than $\sqrt{n}$.

For each such divisor, d, there is another divisor $\dfrac{n}{d}$ greater than $\sqrt{n}$. Their product is $d \times \dfrac{n}{d} = n$, and there are $\tau/2$ such pairs. Then the product of the divisors of n is $n^{\frac{\tau}{2}}$. To illustrate this, let $n = 18$. Then $\tau(18) = 6$ and the 6 divisors are 1, 2, 3, 6, 9, and 18. We group them into $\tau/2$ or 3 pairs: 1 and 18, 2 and 9, and 3 and 6, each of which has product 18. Then the product of all the divisors of 18 is $18^{\frac{6}{2}} = 18^3$.

If n is a square, then τ is odd. Then let $\tau = 2k + 1$. Then there are k divisors less than $\sqrt{n}$. For each such divisor, d, there is another divisor $\frac{n}{d}$, which is greater than $\sqrt{n}$. Their product is n and there are k such pairs. Then there is the divisor $\sqrt{n}$. The product of all the divisors is, therefore, $n^k \sqrt{n} = n^{k+\frac{1}{2}} = n^{\frac{2k+1}{2}} = n^{\frac{\tau}{2}}$. We have proved the following elegant theorem.

Theorem: The product of the divisors of n is $n^{\frac{\tau}{2}}$.   □

Since half of the divisors of n are less than or equal to $\sqrt{n}$, we have an upper bound for $\tau(n)$.

Theorem: $\tau(n) \le 2\sqrt{n}$.   □

This upper bound is quite high. When $n = 100$, for example, it says that $\tau(100) \le 20$, while, in fact, $\tau(100) = \tau(2^2 \times 5^2) = 3 \times 3 = 9$.

THE *SIGMA* FUNCTION

The sum of the divisors of n is denoted $\sigma(n)$. (Lowercase "sigma.") This is often written $\sigma(n) = \sum_{d \mid n} d$, where the summation is over all d such that $d \mid n$, as stated below the sigma sign. The first few values of $\sigma(n)$ are given by $\sigma(1) = 1$, $\sigma(2) = 1 + 2 = 3$, $\sigma(3) = 1 + 3 = 4$, $\sigma(5) = 1 + 5 = 6$, $\sigma(6) = 1 + 2 + 3 + 6 = 12$, $\sigma(7) = 1 + 7 = 8$, and $\sigma(9) = 1 + 3 + 9 = 13$. To find a formula for $\sigma(n)$, we need a few observations. If $n = p^r q^s$, where p and q are prime, then $\sigma(n) = (1 + p + p^2 + \cdots + p^r) \times (1 + q + q^2 + \cdots + q^s)$, since the terms of this product yield all the divisors of n, such as pq^2. This extends to the more general case where $n = p_1^{e_1} p_2^{e_2} \ldots p_r^{e_r}$. Then $\sigma(n) = (1 + p_1 + p_1^2 + \cdots + p_1^{e_1}) \times (1 + p_2 + p_2^2 + \cdots + p_2^{e_2}) \times \cdots \times (1 + p_r + p_r^2 + \cdots + p_r^{e_r})$.

Note that each parenthesis contains a geometric progression. The sum of a geometric progression, $1 + a + a^2 + \cdots + a^m$, is given by $\dfrac{a^{m+1}-1}{a-1}$, as we proved in Chapter 1. Then we have the following expression for $\sigma(n)$:

$$\sigma(n) = \left(\frac{p_1^{e_1+1}-1}{p_1-1}\right) \left(\frac{p_2^{e_2+1}-1}{p_2-1}\right) \cdots \left(\frac{p_r^{e_r+1}-1}{p_r-1}\right) \qquad (6.2)$$

Example: Find $\sigma(100)$. By (6.2), since $100 = 2^2 5^2$, we have $\sigma(100) = \left(\dfrac{2^3-1}{2-1}\right)\left(\dfrac{5^3-1}{5-1}\right) = 7 \times 31 = 217$. This answer can be confirmed by adding $1 + 2 + 4 + 5 + 10 + 20 + 25 + 50 + 100$.

Example: Find $\sigma(2^n)$. By (6.2), we have $\sigma(2^n) = 2^{n+1} - 1$.

Example: Find $\sigma(p)$, where p is prime. $\sigma(p) = 1 + p$, since the divisors of p are 1 and p. Alternatively, we can use (6.2), yielding $\sigma(p) = \dfrac{p^2-1}{p-1} = p + 1$.

MULTIPLICATIVE FUNCTIONS

$\tau(n)$ and $\sigma(n)$ are examples of multiplicative functions. A *multiplicative function, f*, satisfies:

1. $f(1) = 1$.
2. $f(mn) = f(m)f(n)$ whenever $gcd(m,n) = 1$.

If f is a nonidentically zero function satisfying condition 2, it can be easily shown (see the exercises at the end of the chapter) that condition 1 must be true.

That τ and σ are multiplicative follows immediately from (6.1) and (6.2). Other examples of multiplicative functions are the "identity" function, $I(n) = n$, and the constant function, $C(n) = 1$.

PERFECT NUMBERS REVISITED

Recall that a *perfect* number is the sum of its proper divisors. It follows that n is perfect if and only if $\sigma(n) = 2n$, since σ counts n as a divisor.

For example, the perfect number 6 satisfies $\sigma(6) = 1 + 2 + 3 + 6 = 12 = 2 \times 6$. It was stated in Chapter 1 that if $1 + 2 + \cdots + 2^{n-1} = 2^n - 1$ is prime, then $2^{n-1}(2^n - 1)$ is perfect. We shall prove this using the multiplicative property of σ.

Theorem: If $1 + 2 + \cdots + 2^{n-1} = 2^n - 1$ is prime, then $2^{n-1}(2^n - 1)$ is a perfect number.

Proof: Let $p = 2^n - 1$, which we assume is prime. Let $N = 2^{n-1}(2^n - 1) = 2^{n-1}p$. We must show that N is perfect, that is, we must show that $\sigma(N) = 2N$. Here goes. $\sigma(N) = \sigma(2^{n-1}p) = \sigma(2^{n-1})\sigma(p) = (2^n - 1)\sigma(p) = (2^n - 1)(p + 1) = p2^n = 2p2^{n-1} = 2N$, implying that N is perfect. $\qquad \square$

On the other hand, we will show that if an *even* number is perfect, then it must be of the form $2^{n-1}(2^n - 1)$, where $2^n - 1$ is prime. We will make use of the multiplicative nature of σ.

Theorem: If the *even* number N is a perfect number, then $N = 2^{n-1}(2^n - 1)$, where $2^n - 1$ is prime.

Proof: To begin, observe that an even number, N, is of the form $2^{n-1}m$ where m is odd. This follows from the prime decomposition of N. Since 2^{n-1} and m are relatively prime, we have $\sigma(N) = \sigma(2^{n-1}m) = \sigma(2^{n-1})\sigma(m) = (2^n - 1)\sigma(m)$. Furthermore, since N is perfect, $\sigma(N) = 2N = 2^n m$, in which case we have $(2^n - 1)\sigma(m) = 2^n m$. Then $\sigma(m) = 2^n k$, and $m = (2^n - 1)k$. (This is because 2^n and $2^n - 1$ are relatively prime.) Now, if $k > 1$, it would follow that m had the divisors 1, k, and m, implying that $\sigma(m) \geq k + m + 1$. Now, $k + m = k2^n = \sigma(m)$, yielding the impossible equation $\sigma(m) \geq \sigma(m) + 1$. Then $k = 1$, implying that $\sigma(m) = m + 1$, so m is a prime, that is, $2^n - 1$ is indeed a prime. Primes of the form $2^n - 1$ are called *Mersenne primes*. $\qquad \square$

MERSENNE PRIMES

The next theorem supplies a necessary condition for $2^n - 1$ to be prime. This condition is useful in narrowing down the search for Mersenne primes.

Theorem: If $2^n - 1$ is prime, then n must be prime.

Proof: We must show that if n is composite, so is $2^n - 1$. If n is even and greater than 2 (the case $n = 2$ is obvious), then $n = 2k$, with $k \geq 2$, so $2^n - 1 = (2^k)^2 - 1 = (2^k - 1)(2^k + 1)$, which is composite. If n is an odd composite greater than 1, then $n = rs$, where each of r and s is at least 3. The identity $x^m - 1 = (x-1)(x^{m-1} + x^{m-2} + \cdots + x + 1)$ with x replaced by 2^r and m replaced by s yields $2^n - 1 = 2^{rs} - 1 = (2^r)^s - 1 = (2^r - 1)(2^{r(s-1)} + 2^{r(s-2)} + \cdots + 2^r + 1)$, which is composite. □

The fact that n is prime does not guarantee that $2^n - 1$ will be prime. When $n = 11$, for example, $23 \,|\, 2^{11} - 1$.

F(n) = Σf(d) WHERE *d* IS A DIVISOR OF *n*

Given a multiplicative function, f, we define a related function, F, by the formula $F(n) = \sum_{d|n} f(d)$. In other words, $F(n) = f(d_1) + f(d_2) + \cdots + f(d_\tau)$, where $d_1, d_2, \ldots, d_\tau$ are the divisors of n. It is quite interesting that $F(n)$ is also multiplicative, as the next theorem asserts.

Theorem: Let f be multiplicative and let $F(n) = \sum_{d|n} f(d)$. Then F is multiplicative.

Proof: We must show that $F(mn) = F(m)F(n)$ where $gcd(m,n) = 1$. Let the divisors of m be $d_1, d_2, \ldots, d_r$ and let the divisors of n be $d_1', d_2', \ldots, d_s'$. Then the divisors, c, of mn are of the form $d_i d_j'$ where $1 \le i \le r$ and $1 \le j \le s$. Since $gcd(m,n) = 1$, it follows that d_i and d_j' are relatively prime. Then

$$F(mn) = \sum_{c|mn} f(c) = \sum_{d|m, d'|n} f(dd') = \sum_{d|m, d'|n} f(d)f(d') = \sum_{d|m} f(d) \sum_{d'|n} f(d') = F(m)F(n)$$

□

The next to the last equality, $\sum_{d|m, d'|n} f(d)f(d') = \sum_{d|m} f(d) \sum_{d'|n} f(d')$, is hard to verify. It might be easier to write out the two sums on the right and then do the multiplication. Thus, $\sum_{d|m} f(d)$, for example, is $f(d_1) + f(d_2) + \cdots + f(d_r)$, and $\sum_{d'|m} f(d') = f(d_1') + f(d_2') + \cdots + f(d_s')$. Then the product of these sums will yield rs terms of the form dd', where d is one of the r divisors of m and d' is one of the s divisors of n. This is precisely $\sum_{d|m, d'|n} f(d)f(d')$.

$$\tau(n) = \sum_{d|n} C(d) = \sum_{d|n} 1, \text{ while } \sigma(n) = \sum_{d|n} I(d) = \sum_{d|n} d, \text{ so } \tau$$

and σ are multiplicative!

Our next theorem will require manipulating sums. Bear in mind that to each divisor, d, of a given number n, there is a unique divisor $\dfrac{n}{d}$, such that the product of these two divisors is n. As d assumes the values of all the divisors of n in *increasing* order, $\dfrac{n}{d}$ assumes the values of all the divisors of n in *decreasing* order. The following chart demonstrates this for $n = 30$. The product of the entries of each row is, of course, 30.

d	n/d
1	30
2	15
3	10
5	6
6	5
10	3
15	2
30	1

Theorem: Let f and g be multiplicative functions, and let $F(n) = \sum_{d|n} f(d)g\left(\dfrac{n}{d}\right)$. Then $F(n)$ is multiplicative.

Proof: We must show that $F(mn) = F(m)F(n)$ where $gcd(m,n) = 1$. Let the divisors of m be $d_1, d_2, \ldots, d_r$ and let the divisors of n be d_1', $d_2', \ldots, d_s'$. Then the divisors, c, of mn are of the form $d_i d_j'$ where $1 \le i \le r$ and $1 \le j \le s$, and d_i and d_j' are relatively prime.

We have $F(mn) = \sum_{c|mn} f(c)g\left(\dfrac{mn}{c}\right) = \sum_{d|m, d'n} f(dd')g\left(\dfrac{mn}{dd'}\right) = \sum_{d|m, d'n} f(d)$

$f(d')g\left(\dfrac{m}{d}\right)g\left(\dfrac{n}{d'}\right) = \sum_{d|m, d'n} f(d)g\left(\dfrac{m}{d}\right)f(d')g\left(\dfrac{n}{d'}\right) = \sum_{d|m} f(d)g\left(\dfrac{m}{d}\right)\sum_{d'n} f(d')$

$g\left(\dfrac{n}{d'}\right) = F(m)F(n)$, and the proof is over. $\qquad\qquad\square$

If we let g be the constant function, $C(n) = 1$ in the above theorem, we get the result of the previous theorem, that is, $F(n) = \sum_{d|n} f(d)$ is multiplicative.

The values of a multiplicative function, f, are completely determined by its values for integers of the form p^k, where p is a prime, since $n = p_1^{e_1} p_2^{e_2} \ldots p_r^{e_r}$ implies that $f(n) = f(p_1^{e_1} p_2^{e_2} \ldots p_r^{e_r}) = f(p_1^{e_1}) f(p_2^{e_2}) \ldots f(p_r^{e_r})$.

Example: Obtain formula (6.1) for $\tau(n)$ using the fact that τ is multiplicative. Since p^k has the $k + 1$ divisors $1, p, p^2, \ldots, p^k$, we have $\tau(p^k) = k + 1$, from which formula (6.1) follows at once.

Example: Show that $\sum_{d|n} \dfrac{1}{d} = \dfrac{\sigma(n)}{n}$. We have $\sigma(n) = \sum_{d|n} d = \sum_{d|n} \dfrac{n}{d} = n \sum_{d|n} \dfrac{1}{d}$, from which the desired result follows by transposing the n.

THE MÖBIUS FUNCTION

The so-called Möbius function, denoted $\mu(n)$, is defined as follows:

1. $\mu(1) = 1$.
2. $\mu(n) = 0$ if $p^2 \mid n$ for any prime p.
3. $\mu(n) = (-1)^r$ if $n = p_1 p_2 \ldots p_r$.

The first few values of $\mu(n)$ are $\mu(1) = 1, \mu(2) = -1, \mu(3) = -1, \mu(4) = 0$ (since $2^2 \mid 4$), $\mu(5) = -1, \mu(6) = 1, \mu(7) = -1, \mu(8) = 0$ (since $2^2 \mid 8$), $\mu(9) = 0$ (since $3^2 \mid 9$), and $\mu(10) = 1$. If $n > 1$, it must be square-free for $\mu(n)$ to be nonzero. (A number is called *square-free* if its only square divisor is 1. Thus, for example, 10 is square-free, while 18 is not square-free since $9 \mid 18$.) This is why #3 above assumes that $n = p_1 p_2 \ldots p_r$.

Theorem: $\mu(n)$ is multiplicative.

Proof: We must show that if $gcd(m,n) = 1$, then $\mu(mn) = \mu(m)\mu(n)$. If either of m and n equals 1, this is obvious since $\mu(1) = 1$. Furthermore, if either of m and n contains a divisor of the form p^2, where p is prime, this is also obvious since we would have $p^2 \mid mn$, implying that $\mu(mn) = \mu(m)\mu(n) = 0$. Let us assume, then, that m and n are square-free and are greater than 1. Then $m = p_1 p_2 \ldots p_r$ and $n = q_1 q_2 \ldots q_s$. Then $mn = p_1 p_2 \ldots p_r q_1 q_2 \ldots q_s$, in which case $\mu(mn) = (-1)^{r+s} = (-1)^r (-1)^s = \mu(m)\mu(n)$. $\square$

Here is a theorem about μ. We present two proofs, each as interesting as the theorem.

Theorem: Let $g(n) = \sum_{d|n} \mu(d)$. Then $g(n) = \begin{cases} 1 & n=1 \\ 0 & n>1 \end{cases}$.

Proof 1: If $n=1$, then $g(n) = \mu(1) = 1$, as the theorem states. If, on the other hand, $n>1$, then assume that $n = p_1^{e_1} p_2^{e_2} \ldots p_r^{e_r}$. Now, since $\mu(d) = 0$ unless d is a square-free divisor of n, it suffices to evaluate the sum for the divisor 1; divisors of the form p_i (i.e., divisors consisting of a single prime), $p_i p_j$ (divisors consisting of a product of two distinct primes), $p_i p_j p_k$ (divisors consisting of a product of three distinct primes), and so on; and, finally, the divisor $p_1 p_2 \ldots p_r$.

Note that since n has an r distinct prime divisors, there are $\binom{r}{1}$ divisors of the form p_i (for which $\mu = -1$), $\binom{r}{2}$ divisors of the form $p_i p_j$ (for which $\mu = 1$), $\binom{r}{3}$ divisors of the form $p_i p_j p_k$ (for which $\mu = -1$), and so on and, finally, $\binom{r}{r}$ divisors of the form $p_1 p_2 \ldots p_r$ (for which $\mu = (-1)^r$). Furthermore, there are $\binom{r}{0}$ divisors with no prime factors, namely, 1 (for which $\mu = 1$). Then $\mu(n) = \binom{r}{0} - \binom{r}{1} + \binom{r}{2} - \binom{r}{3} + \cdots (-1)^r \binom{r}{r}$, which, by an identity in Chapter 3, equals 0, thereby completing the proof. □

Proof 2: Since $g(n)$ is multiplicative, it suffices to show that the theorem is true when $n = p^k$. We then have $g(p^k) = \mu(1) + \mu(p) + \mu(p^2) + \cdots + \mu(p^k) = 1 + (-1) = 0$. □

One of the reasons that μ is important in number theory is given by the next theorem, often called the *Möbius inversion formula*. We omit the proof.

Theorem: Let $g(n) = \sum_{d|n} f(d)$. Then $f(n) = \sum_{d|n} g(d) \mu\left(\frac{n}{d}\right) = \sum_{d|n} g\left(\frac{n}{d}\right) \mu(d)$. □

Letting $f(d) = C(d) = 1$, we know that $g(n) = \tau(n)$, as noted earlier in the chapter. Then we obtain the beautiful identity

$$1 = \sum_{d|n} \tau(d)\mu\left(\frac{n}{d}\right)$$

Example: When $n = 10$, this becomes $1 = \tau(1)\mu(10) + \tau(2)\mu(5) + \tau(5)\mu(2) + \tau(10)\mu(1) = \tau(1) - \tau(2) - \tau(5) + \tau(10) = 1 - 2 - 2 + 4$, which is correct.

Letting $f(d) = I(d) = d$, we obtain $g(n) = \sigma(n)$, in which case we have

$$n = \sum_{d|n} \sigma(d)\mu\left(\frac{n}{d}\right)$$

Example: When $n = 10$, as in the previous example, this becomes $10 = \sigma(1)\mu(10) + \sigma(2)\mu(5) + \sigma(5)\mu(2) + \sigma(10)\mu(1) = \sigma(1) - \sigma(2) - \sigma(5) + \sigma(10) = 1 - 3 - 6 + 18$, which is, once again, correct.

THE RIEMANN ZETA FUNCTION

The Riemann zeta function is defined by $\zeta(s) = \sum_{n=1}^{\infty} \frac{1}{n^s} = 1 + \frac{1}{2^s} + \frac{1}{3^s} + \frac{1}{4^s} + \cdots$ where s is a complex number with certain restrictions. This function is very important in a number of branches of mathematics and has been studied extensively. In this elementary text, we will restrict s to positive integer values and will only get a tiny glimpse of this deep and mysterious function. When $s = 1$, we obtain a *divergent* series, that is, a series whose partial sums get larger and larger without bound. The partial sums approach infinity, as we will see. Formally, $\zeta(1) = \sum_{n=1}^{\infty} \frac{1}{n} = 1 + \frac{1}{2} + \frac{1}{3} + \frac{1}{4} + \cdots$. Now, observe the following inequalities:

$$\frac{1}{2} \geq \frac{1}{2}$$

$$\frac{1}{3} + \frac{1}{4} \geq \frac{1}{4} + \frac{1}{4} \geq \frac{1}{2}$$

$$\frac{1}{5} + \frac{1}{6} + \frac{1}{7} + \frac{1}{8} \geq \frac{1}{8} + \frac{1}{8} + \frac{1}{8} + \frac{1}{8} \geq \frac{1}{2}$$

$$\frac{1}{9}+\frac{1}{10}+\frac{1}{11}+\frac{1}{12}+\frac{1}{13}+\frac{1}{14}+\frac{1}{15}+\frac{1}{16}\geq\frac{1}{16}+\frac{1}{16}+\frac{1}{16}$$

$$+\frac{1}{16}+\frac{1}{16}+\frac{1}{16}+\frac{1}{16}+\frac{1}{16}\geq\frac{1}{2}$$

Continuing in this manner, we see that a partial sum of $\zeta(1)$ can be made as large as we please by accumulating enough consecutive sequences of terms that exceed $\frac{1}{2}$. The series $1+\frac{1}{2}+\frac{1}{3}+\frac{1}{4}+\cdots$ is called the *harmonic series*, and its divergence to infinity was known in the fourteenth century.

The eighteenth-century mathematician Euler showed, using calculus, that $\zeta(2)=\frac{\pi^2}{6}$, that is,

$$\sum_{n=1}^{\infty}\frac{1}{n^2}=1+\frac{1}{2^2}+\frac{1}{3^2}+\frac{1}{4^2}+\frac{1}{5^2}+\cdots=\frac{\pi^2}{6} \tag{6.3}$$

Consider the infinite product $P=\left(1-\frac{1}{2^2}\right)\left(1-\frac{1}{3^2}\right)\left(1-\frac{1}{5^2}\right)$ $\left(1-\frac{1}{7^2}\right)\ldots$, where the factors are of the form $\left(1-\frac{1}{p^2}\right)$, as p ranges over all the primes. Let's calculate $\zeta(2)\times P$. We will do this by multiplying by one factor of P at a time:

$$\left(1+\frac{1}{2^2}+\frac{1}{3^2}+\frac{1}{4^2}+\frac{1}{5^2}+\frac{1}{6^2}+\frac{1}{7^2}+\frac{1}{8^2}+\cdots\right)\left(1-\frac{1}{2^2}\right)$$

$$=\left(1+\frac{1}{3^2}+\frac{1}{5^2}+\frac{1}{7^2}+\frac{1}{9^2}+\cdots\right)$$

Thus, multiplication by $\left(1-\frac{1}{2^2}\right)$ removes every term whose denominator is divisible by 2. Now, multiply the right side of this equation by the next factor, $\left(1-\frac{1}{3^2}\right)$:

$$\left(1+\frac{1}{3^2}+\frac{1}{5^2}+\frac{1}{7^2}+\frac{1}{9^2}+\cdots\right)\left(1-\frac{1}{3^2}\right)$$

$$=\left(1+\frac{1}{5^2}+\frac{1}{7^2}+\frac{1}{11^2}+\frac{1}{13^2}+\cdots\right)$$

Multiplication by $\left(1-\dfrac{1}{3^2}\right)$ removes every term whose denominator is divisible by 3. Now, we multiply the right side of this equation by the next factor, $\left(1-\dfrac{1}{5^2}\right)$:

$$\left(1+\frac{1}{5^2}+\frac{1}{7^2}+\frac{1}{11^2}+\frac{1}{13^2}+\cdots\right)\left(1-\frac{1}{5^2}\right)$$

$$=\left(1+\frac{1}{7^2}+\frac{1}{11^2}+\frac{1}{13^2}+\frac{1}{17^2}+\cdots\right)$$

Multiplication by $\left(1-\dfrac{1}{5^2}\right)$ removes every term whose denominator is divisible by 5. As we continue this multiplication, we remove more terms. In fact, each term except 1 eventually gets removed. We conclude that $\zeta(2) \times P = 1$. Using (6.3), we then see that $P = \dfrac{1}{\varsigma(2)} = \dfrac{6}{\pi^2}$ (approximately .608). We are now ready to state and prove a beautiful theorem.

Theorem: The probability that two randomly chosen positive integers are relatively prime is $\dfrac{6}{\pi^2}$.

Proof: $\dfrac{1}{n}$ of all positive integers are divisible by n, since every nth positive integer is a multiple of n. When $n = 3$, for example, this says that every third positive integer is a multiple of 3, that is, 3 divides $\dfrac{1}{3}$ of all positive integers. Then the probability that a given integer is divisible by n is $\dfrac{1}{n}$. Now, if we choose two integers, the probability that both are divisible by n is the product of the individual probabilities, that is, $\dfrac{1}{n^2}$. It follows that the probability that n is not a common divisor is $1-\dfrac{1}{n^2}$.

To finish the proof, observe that two positive integers are relatively prime if they share no prime divisors. Since these events are independent, we multiply the probabilities of this for each prime. But this is exactly P, so the proof is complete. □

We shall not prove it in this text, but $\dfrac{6}{\pi^2}$ is also the probability that a randomly chosen positive integer is square-free.

Furthermore, the probability that three randomly chosen integers do not share a common divisor is the reciprocal of $\zeta(3)$ (approximately .832).

EXERCISES

An asterisk (∗) indicates that the exercise can be developed into a research project.

1   Find $\tau(300)$ by finding all the divisors of 300 and counting them. Then use the formula. Do the same for $\tau(1000)$. *See the programs at the end of the chapter.*

2   Describe all numbers n, such that $\tau(n) = 6$. What is the smallest such n?

3   Describe all numbers n, such that $\tau(n) = 12$. What is the smallest such n?

4   ∗Given any positive integer m, show that there exists a positive integer, n, such that $\tau(n) = m$.

5   ∗Show that if $m \geq 2$ in the previous exercise, then there exist infinitely many n's with $\tau(n) = m$.

6   Find $\sigma(1000)$ by adding all the divisors of 1000. Then use the formula.

7   Find $\sigma(p^3)$, where p is prime, (i) directly from the definition and (ii) using formula (6.2). Reconcile your answers.

8   Show that if $n = 2^k$, where $k \geq 1$, then $\sigma(n) = 2n - 1$.

9   Assuming that a function, f, such that $f(k) \neq 0$ for all $k \geq 1$, satisfies $f(mn) = f(m)f(n)$ whenever $gcd(m,n) = 1$, show that $f(1) = 1$.

10   How does it follow from formulas (6.1) and (6.2) that τ and σ are multiplicative?

11   Show that the identity function, $I(n) = n$, and the constant function, $C(n) = 1$, are multiplicative.

12   Show that the function $f(n) = n^k$, where k is a nonnegative integer, is multiplicative. How does this apply to the previous exercise?

13   If f and g are multiplicative, show that fg and f/g (if f/g is defined) are also multiplicative.

14 *Let f be a multiplicative function that is not identically zero. Show that $\sum_{d|n} \mu(d)f(d) = (1-f(p_1))(1-f(p_2))\ldots(1-f(p_r))$, where p_1, p_2, ..., p_r are the prime divisors of n.

15 *Let $n = p_1^{e_1} p_2^{e_2} \ldots p_r^{e_r}$. Show using the previous exercise that:

   a. $\displaystyle\sum_{d|n} \mu(d)\tau(d) = (-1)^r$.

   b. $\displaystyle\sum_{d|n} \mu(d)\sigma(d) = (-1)^r p_1 p_2 \ldots p_r$.

   c. $\displaystyle\sum_{d|n} \mu(d)d = (1-p_1)(1-p_2)\ldots(1-p_r)$.

   d. $\displaystyle\sum_{d|n} \frac{\mu(d)}{d} = \left(1-\frac{1}{p_1}\right)\left(1-\frac{1}{p_2}\right)\ldots\left(1-\frac{1}{p_r}\right)$.

16 If n is a perfect number, show that $\displaystyle\sum_{d|n} \frac{1}{d} = 2$.

17 Show that 6 is the only square-free even perfect number.

18 Show that p^k, where p is prime, is not a perfect number.

19 If m is an even perfect number greater than 6, show that $m = 4 \ (mod\ 6)$.

20 The *harmonic mean*, H, of the n numbers $a_1, a_2, \ldots, a_n$ is defined by the equation $\displaystyle\frac{1}{H} = \frac{1}{n}\left(\frac{1}{a_1} + \frac{1}{a_2} + \cdots + \frac{1}{a_n}\right)$. Write a program to obtain the harmonic mean of any given set of numbers. Find the harmonic mean of various sets of consecutive integers and try to develop an idea of where it falls.

21 Show that the harmonic mean of the divisors of an even perfect number is an integer.

22 The *geometric mean* of the n numbers $a_1, a_2, \ldots, a_n$ is $\sqrt[n]{a_1 a_2 \ldots a_n}$. Write a program to obtain the geometric mean of any given set of numbers. Compare it to the harmonic mean after computing several examples of both for various sets of integers.

23 *A number is called *multiplicatively perfect* if it is the product of its proper divisors. If p is a prime, show that p^3 is multiplicatively perfect. Are there other examples less than 1000?

24 *Show that semiprimes are multiplicatively perfect. (A *semiprime* is the product of two primes.)

25 $^*r$ and s are *amicable numbers* if each is the sum of the proper divisors of the other, that is, if $r = \sigma(s) - s$ and $s = \sigma(r) - r$. Let $a = 3 \times 2^n - 1$, $b = 3 \times 2^{n-1} - 1$, and $c = 9 \times 2^{2n-1} - 1$, where $n \geq 2$. If a, b, and c are prime, show that $2^n ab$ and $2^n c$ are amicable numbers. Note that when $n = 2$, we get the amicable pair 220 and 284, known to the Pythagoreans.

26 Find $\mu(n)$ for all $n \leq 200$.

27 *Show that $\mu(n)\mu(n+1)\mu(n+2)\mu(n+3) = 0$ for all $n \geq 1$.

28 Find $\sum_{k=1}^{1000} \mu(k!)$.

29 $^*\omega(n)$ is the number of *distinct* prime divisors of n. Also, $\omega(1) = 0$. Show that if $gcd(m,n) = 1$, then $\omega(mn) = \omega(m) + \omega(n)$. Use this to show that $2^{\omega(n)}$ is multiplicative.

30 Show that $\sum_{d|n} |\mu(d)| = 2^{\omega(n)}$.

31 *Show that $\sum_{d|n} |\mu(d)|$ counts the number of square-free divisors of n.

32 Show that every positive integer can be written as mn^2, where m is square-free. Write a program to do this.

33 *Let $r \geq 0$. Define $\sigma_r(n) = \sum_{d|n} d^r$. Show that $\sigma_r(n)$ is multiplicative.

34 Show that $\tau(n) = \sigma_0(n)$ and $\sigma(n) = \sigma_1(n)$.

35 Show that $\sigma_r(p^k) = \dfrac{p^{(k+1)r} - 1}{p^r - 1}$, where p is a prime.

36 Let f be a multiplicative function such that $\mu(n) = \sum_{d|n} f(d)$. Find $f(n)$.

37 Let f and g be multiplicative functions such that $f(p^k) = g(p^k)$ for all primes, p, and for all $k \geq 1$. Show that $f(n) = g(n)$ for all n.

38 Prove that $\sum_{d|n} \sigma(d) = \sum_{d|n} \tau(d)\left(\dfrac{n}{d}\right)$.

39 *Ramanujan called a number n *highly composite* if $\tau(k) < \tau(n)$ for all $k < n$. The first few highly composite numbers are 2, 4, 6, and 12. Find the next 100 such numbers.

40 *It is known that $\left(\sum_{d|n} \tau(d)\right)^2 = \sum_{d|n} \tau(d)^3$. Verify this formula for $n \le 20$.

41 By letting $n = p^{k-1}$, where p is prime, show that the identity $1^3 + 2^3 + 3^3 + \cdots + k^3 = (1 + 2 + 3 + \cdots + k)^2$ is a special case of the remarkable formula of the previous exercise.

42 *Let $P(s) = \left(1 - \dfrac{1}{2^s}\right)\left(1 - \dfrac{1}{3^s}\right)\left(1 - \dfrac{1}{5^s}\right)\left(1 - \dfrac{1}{7^s}\right)\cdots$. Show that $\zeta(s) = \dfrac{1}{P(s)}$.

43 *Let $s \ge 2$. Show that $\dfrac{1}{\zeta(s)} = \sum_{n=1}^{\infty} \dfrac{\mu(n)}{n^s}$.

44 *Prove that there are infinitely many pairs of consecutive square-free numbers as follows:

    a. Assume to the contrary that there are only finitely many pairs of square-free numbers. Why would it follow that there is an integer N such that there are no pairs of consecutive square-free numbers larger than N?

    b. As a consequence of the preceding assumption, show that it would follow that at least half of all numbers larger than N are not square-free.

    c. Show that this would contradict the fact (stated without proof in the chapter) that the probability that a randomly chosen positive integer is square-free is $\dfrac{6}{\pi^2}$.

45 Write a program to find all pairs of consecutive square-free numbers up to any given number n.

46 *Given the consecutive integers a and $a + 1$, there are infinitely many pairs, x and $x + 1$, such that $a \mid x$ and $a + 1 \mid x + 1$. When $a = 3$, for example, the integers $x = 15$ and $x + 1 = 16$ have the property that $3 \mid x$ and $4 \mid x + 1$. Write a program to find arbitrarily many solutions, x and $x + 1$, for a given pair of consecutive integers a and $a + 1$.

47 *Find the smallest positive integer, n, such that $\tau(n) = m$, where m is a given positive integer.

48 *Note that $\sigma(3) = 1 + 3 = 4 = 2^2$ and $\sigma(22) = 1 + 2 + 11 + 22 = 36 = 6^2$. Find all numbers $n \le 2000$ for which $\sigma(n)$ is a square.

49 *A perfect number, n, satisfies $\sigma(n) = 2n$. This has been generalized to *multiply perfect* numbers satisfying $\sigma(n) = kn$ where $k > 2$. Find all $n \leq 1000$ for which $\sigma(n) = 3n$.

50 *The number 12 has the curious property that the *product* of its proper divisors equals its square, that is, $1 \times 2 \times 3 \times 4 \times 6 = 144 = 12^2$. Find all $n \leq 100$ with this property.

51 Verify that 9,363,584 and 9,437,056 are amicable numbers, that is, verify that each is the sum of the proper divisors of the other.

52 *Let $f(n)$ be the sum of the *proper* divisors of n. Furthermore, define $f^k(n)$ as follows. Let $f^1(n) = f(n)$, $f^2(n) = f[f(n)]$, $f^3(n) = f[f[f(n)]]$, etc. Show that when $n = 12,496$, we have $f^5(12,496) = n = 12,496$, thereby forming a so-called amicable chain of length 5.

53 *If $n = 14,316$ in the previous exercise, show that we obtain an amicable chain of length 28, that is, show that $f^{28}(14,316) = n = 14,316$.

54 *Let $f(n) = \begin{cases} 3n+1 & n \text{ odd} \\ \dfrac{n}{2} & n \text{ even} \end{cases}$. Define $f^k(n)$ as in the two preceding exercises. It is conjectured that given n, there exists a k such that $f^k(n) = 1$. Verify this for $n \leq 1000$. Note that $f(1) = 2$ and $f(2) = 1$. The sequence $n, f^1(n), f^2(n), f^3(n), \ldots$ is the *orbit* of n under f. The orbit of 1 under f is 1, 2, 1, 2, ... An orbit consisting of a repeated sequence of values is *periodic*. The conjecture says that the orbit of any n gets "pulled" into the periodic orbit 1, 2, 1, 2, ...

55 *Let $f(n) = \begin{cases} \left\lfloor n^{\frac{3}{2}} \right\rfloor & n \text{ odd} \\ \lfloor \sqrt{n} \rfloor & n \text{ even} \end{cases}$. It has been conjectured that given any positive integer n, there exists a k such that $f^k(n) = 1$. Verify this for all $n \leq 100$. Given an n, its orbit under the action of this function is called a *juggler sequence*.

56 Write a program to compute the sum of the first k terms of $\zeta(n)$ for given positive integers n and k. Show that your answer for $\zeta(4)$ approaches $\pi^4/90$ as k gets large. It is known that when n is even, $\zeta(n)$ is a rational multiple of π^n. Not much is known about $\zeta(n)$ when n is odd.

57 *Observe that the consecutive triple, 33, 34, and 35, consists of semiprimes, that is, each member of the triple is the product of two primes. Find the next four consecutive triples with this property.

Stand-Alone Program Listing the Proper Factors

```
<!DOCTYPE HTML>
<html>
<head>
<meta charset="UTF-8">
<head>
<title>Return proper factors</title>
<script>

function produceFactors(n) {
  var ans = []; //start off with empty array
  for (var i=1;i<n;i++) {
    if ((n%i)==0) {
      ans.push(i);
    }
  }
  return ans;
}
function getfactors() {
  placeref = document.getElementById("place");
  messages = "The proper factors are <br/>";
  n = parseInt(document.f.num.value);
  factors = produceFactors(n);
  for (var i=0;i<factors.length;i++) {
    messages +=String(factors[i])+"<br/>";
  }
  placeref.innerHTML = messages;
  return false;
}
</script>
</head>
<body>
Produce list of proper factors:
<form name="f" onsubmit="return getfactors();">
 <input type="number" name="num" value=""/>
 <input type="submit" value="Enter integer"/>
```

```
</form>
<div id="place">
List of factors will go here.
</div>
</body>
</html>
```

Tau from the Definition

```
<!DOCTYPE HTML>
<html>
<meta charset="UTF-8">
<head>
<title>Tau</title>
<script>
function produceFactors(n){
  var ans = []; //start off with empty array
  for (var i=1;i<=n;i++){
    if ((n%i)==0){
       ans.push(i);
    }
  }
  return ans;
}
function tau(){
  placeref = document.getElementById("place");
  n = parseInt(document.f.num.value);
  t = produceFactors(n).length;
  messages = "The value of tau for "+ String(n)+" is "+
String(t);
  placeref.innerHTML = messages;
  return false;
}
</script>
</head>
<body>
Enter number to see its tau value:
<form name="f" onSubmit="return tau();">
 <input type="number" name="num" value=""/>
 <input type="submit" value="Enter integer"/>
</form>
<div id="place">
```

```
Answer will go here.
</div>
</body>
</html>
```

Prime Decomposition File to Be Included Using External Script Element

```javascript
// JavaScript

function produceFactors(n) {
  var ans = []; //start off with empty array
  //notice start with 2, not 1
  for (var i=2;i<=n;i++) {
    if ((n%i)==0) {
      ans.push(i);

    }
  }

  return ans;
}
function onlyprimes(factors) {
  var ans = [];
  for (var i=0;i<factors.length;i++) {
    if (isPrime(factors[i])) {
      ans.push(factors[i]);
    }
  }
  return ans;
}
function buildPrimeDecomposition(n) {
  var table = []; //each element will be an array of
  length 2
  var m = n;
  var facs = produceFactors(n);
    while (facs.length>0) {
      f = facs[0]; //always using first, that is, 0th
      //element
      if ((m % f )>0) {
          //f, the ith element of facs does NOT
          //divide m
```

```javascript
        facs.shift();  //remove this factor
      }
      else {
        m = m / f;
        e=1;
        //need to see if exponent is more than 1
        while (0 == (m%f ) ) {
          m = m /f;
          e++;
        }
        table.push ([f,e]);
        facs.shift();
      }


    }
    return table;
}
function isPrime(n) {
    var lim = Math.sqrt(n);
    for (var i=2;i<=lim;i++) {
      if (0 == n % i) {
        return false;
      }
    }
    return true;
}
```

Tau from Formula

```html
<!DOCTYPE HTML>
<html>
<meta charset="UTF-8">
<head>
<title>Tau from formula</title>
<script type="text/javascript"
src="primeDecomposition.js"> </script>
<script>
function tauFF() {
  var table;
  placeref = document.getElementById("place");
  n = parseInt(document.f.num.value);
  table = buildPrimeDecomposition(n);
```

```
  t = 1;
  for (var i=0;i<table.length;i++){
    t = t * (1+table[i][1]);
  }
  messages = "The value of tau for "+ String(n)+" is "+
String(t);
  placeref.innerHTML = messages;
  return false;
}
</script>
</head>
<body>
Enter number to see its tau value:
<form name="f" onSubmit="return tauFF();">
 <input type="number" name="num" value=""/>
 <input type="submit" value="Enter integer"/>
</form>
<div id="place">
Answer will go here.
</div>
</body>
</html>
```

7

THE EULER PHI FUNCTION

The so-called Phi function, developed by the great Swiss mathematician, Leonard Euler, is involved in many theorems of number theory and other branches of mathematics.

THE *PHI* FUNCTION

The number of positive integers less than n that are relatively prime to n is denoted $\phi(n)$. The first few values of this important function, called *the Euler ϕ function*, are given by $\phi(1) = 1$, $\phi(2) = 1$, $\phi(3) = 2$, $\phi(4) = 2$, $\phi(5) = 4$, $\phi(6) = 2$ (since only 1 and 5 are relatively prime to 6), $\phi(7) = 6$, $\phi(8) = 4$, $\phi(9) = 6$, and $\phi(10) = 4$.

Computing the Euler phi function from the definition can be done using a *gcd* function. I started with the function from a previous chapter and removed the display of messages. The document has three functions: find for the interaction with the user, gcd for returning the greatest common divisor, and phi for checking each number from 1 to the specified number n. The find function is invoked by action of the form. The find function invokes the

Elementary Number Theory with Programming, First Edition. Marty Lewinter and Jeanine Meyer.
© 2016 John Wiley & Sons, Inc. Published 2016 by John Wiley & Sons, Inc.

phi function, which invokes the gcd function. The phi function contains a for-loop and checks and computes the *gcd* value of the loop variable and n. If the value is one, a variable named ans is incremented. When the loop is complete, the ans value is returned. The find function displays the results.

Observe that if p is a prime, $\phi(p) = p - 1$. To calculate $\phi(p^k)$, notice that the numbers less than or equal to p^k that are *not* relatively prime to it are the p^{k-1} multiples of p, namely, $p, 2p, 3p, \ldots, p^{k-1}p$. Since there are p^k positive integers less than or equal to p^k, subtraction yields

$\phi(p^k) = p^k - p^{k-1} = p^k\left(1 - \dfrac{1}{p}\right)$. When $p = 2$, this yields $\phi(2^k) = 2^k(\frac{1}{2}) = 2^{k-1}$, which is in perfect agreement with our earlier observation that $\phi(8) = 4$.

Observe, also, that $\phi(n)$ is *even* when $n > 1$. To see this, observe that given a such that $1 \le a \le n$, then $gcd(a,n) = 1$ if and only if $gcd(n - a, n) = 1$, since any divisor of a and n must also divide $n - a$, while any divisor of $n - a$ and n must also divide their sum, a. (For example, both 7 and 13 are relatively prime to 20, while neither of 5 and 15 is.)

We claim that ϕ is multiplicative, that is, if a and b satisfy $gcd(a,b) = 1$, then $\phi(ab) = \phi(a)\phi(b)$. The proof is difficult and we will skip it.

Theorem: ϕ is multiplicative, that is, given a and b such that $gcd(a,b) = 1$, then $\phi(ab) = \phi(a)\phi(b)$. Then these $\phi(a)\phi(b)$ numbers are relatively prime to both a and b, implying that they are relatively prime to ab. So $\phi(ab) = \phi(a)\phi(b)$.   □

If $n = p_1^{e_1} p_2^{e_2} \ldots p_r^{e_r}$, we have, since ϕ is multiplicative, $\phi(n) = \phi\left(p_1^{e_1}\right)$

$\phi\left(p_2^{e_2}\right) \ldots \phi\left(p_r^{e_r}\right) = p_1^{e_1}\left(1 - \dfrac{1}{p_1}\right) p_2^{e_2}\left(1 - \dfrac{1}{p_2}\right) \ldots p_r^{e_r}\left(1 - \dfrac{1}{p_r}\right) = n\left(1 - \dfrac{1}{p_1}\right)$

$\left(1 - \dfrac{1}{p_2}\right) \ldots \left(1 - \dfrac{1}{p_r}\right)$, which yields the elegant formula

$$\phi(n) = n\left(1 - \frac{1}{p_1}\right)\left(1 - \frac{1}{p_2}\right) \ldots \left(1 - \frac{1}{p_r}\right) \tag{7.1}$$

Example: To find $\phi(100)$, note that the prime divisors are 2 and 5. It follows, using (7.1), that $\phi(100) = 100(1 - 1/2)(1 - 1/5) = 100(1/2)(4/5) = 40$.

Creating a program to compute the Euler phi function from formula (7.1) is similar to what was shown in the last chapter for the tau function. We use an external script element to bring in the code containing the `buildPrimeDecomposition` function. Recall that this produces an array of arrays, with the first element of each of the inner arrays holding a prime number and the second element holding the exponent. For the phi function, we only need the primes. The critical part of the code is:

```
table = buildPrimeDecomposition(n);
  t = n;
  for (var i=0;i<table.length;i++){
    t = t * (1-(1/table[i][0]));
  }
```

However, an ugly problem arose that is a result of the calculations involving floating point numbers. The answer produced for 9 was 6.0000000001. So, knowing that the answer really is an integer, I added the line:

```
t = Math.floor(t);
```

However, it really made me favor the definition over the formula.

Example: Show that if n is odd, then $\phi(2n) = \phi(n)$. Since ϕ is multiplicative, and since $gcd(2,n) = 1$, we have $\phi(2n) = \phi(2)\phi(n) = \phi(n)$.

Let $f(n) = \sum_{d|n} \phi(d)$. Then f is multiplicative, since ϕ is. To evaluate $f(n)$, assume first that $n = p^k$, where p is prime. Then we have $f(p^k) = \phi(1) + \phi(p) + \phi(p^2) + \phi(p^3) + \cdots + \phi(p^k) = 1 + (p-1) + (p^2 - p) + (p^3 - p^2) + \cdots + (p^k - p^{k-1}) = p^k$, from which it follows that $f(n) = n$. This yields the identity

$$n = \sum_{d|n} \phi(d) \tag{7.2}$$

Example: Verify (7.2) for $n = 10$. We have $10 = \phi(1) + \phi(2) + \phi(5) + \phi(10) = 1 + 1 + 4 + 4$.

Here is another proof of (7.2), which we will present for the special case $n = 12$ but whose generality will be apparent. Consider the 12 fractions 1/12, 2/12, 3/12, 4/12, 5/12, 6/12, 7/12, 8/12, 9/12, 10/12, 11/12, and 12/12. When we reduce these 12 fractions to lowest terms, the denominators will be the divisors of 12, namely, 1, 2, 3, 4, 6, and 12. The number of fractions for a divisor, d, of 12 will be precisely $\phi(d)$. Thus, for $d = 4$ (for which $\phi(4) = 2$), we get the 2 reduced fractions 1/4 and 3/4 from 3/12 and 9/12, respectively, while for $d = 6$ (for which $\phi(6) = 2$), we get the 2 reduced fractions 1/6 and 5/6 from 2/12 and 10/12, respectively.

Applying Möbius inversion to (7.2) yields the identity

$$\phi(n) = \sum_{d|n} d \cdot \mu\left(\frac{n}{d}\right) = \sum_{d|n} \frac{n}{d} \mu(d) = n \sum_{d|n} \frac{\mu(d)}{d}$$

or more to the point,

$$\phi(n) = n \sum_{d|n} \frac{\mu(d)}{d} \qquad (7.3)$$

How do we reconcile formulas (7.1) and (7.3)? It would require equating the right sides of both formulas, canceling the n and then showing that the resulting equation is true, that is, we must show that

$$\left(1 - \frac{1}{p_1}\right)\left(1 - \frac{1}{p_2}\right)\cdots\left(1 - \frac{1}{p_r}\right) = \sum_{d|n} \frac{\mu(d)}{d}$$

Now, since $\mu(d) = 0$ unless d is square-free, the only nonvanishing terms on the right will result when $d = p_i$, $p_i p_j$, $p_i p_j p_k$, etc., that is, when d is the product of some (or all) of the primes $p_1, p_2, \ldots, p_r$, each of which is raised to the power 1. Now, $\mu(d)$ will be either 1 or -1, depending on the parity of the number of primes in d. Thus, the sum on the right will agree with the sum on the left since they will yield precisely the same terms. For example, the term $\dfrac{-1}{p_1 p_2 p_3}$, occurring on the left, will certainly occur on the right since $\mu(p_1 p_2 p_3) = -1$, with an analogous statement concerning a term such as $\dfrac{1}{p_1 p_2 p_4 p_7}$.

Let's add up all the numbers less than n and relatively prime to it. If a is such a number, so is $n - a$, as we observed earlier in the chapter. There are $\phi(n)/2$ such pairs, each with sum n. It follows that the sum of all the numbers less than n and relatively prime to it is $n\phi(n)/2$. Stated formally, we have

$$\sum_{k \leq n, \, \gcd(k,n)=1} k = \frac{n\phi(n)}{2}$$

Example: The sum of all the numbers less than 10 and relatively prime to it is $10 \times \phi(10)/2$, that is, $1 + 3 + 7 + 9 = 10 \times 4/2 = 20$.

EULER'S GENERALIZATION OF FERMAT'S LITTLE THEOREM

Recall the theorem of Chapter 5 that says that $ax = b$ (*mod n*) is solvable if and only if $gcd(a,n) \mid b$, and then consider the modular equation $ax = 1$ (*mod n*). It follows that this is solvable *if and only if gcd(a,n) = 1*. We then see that a least residue, a (*mod n*), has a multiplicative inverse if and only if $gcd(a,n) = 1$. Then there are exactly $\phi(n)$ such residues. If we multiply each of the members of this set of invertible numbers by any one of them, we must get a rearrangement of the set. (For example, the four invertible residues of $n = 10$ are 1, 3, 7, and 9. If we multiply each of them by, say, 3, we get 3, 9, 21, and 27, which, when reduced *mod* 10, yield 3, 9, 1, and 7, which is a rearrangement of the original set, 1, 3, 7, 9.)

Using an argument analogous to the one Fermat used to get his Little Theorem, Euler reasoned as follows. Let the invertible residues of n be $a_1, a_2, \ldots, a_r$, where $r = \phi(n)$. Then let $gcd(a,n) = 1$ in which case $(aa_1)(aa_2)\ldots(aa_r) = a_1a_2\ldots a_r$ (*mod n*). This becomes $a^r = 1$ (*mod n*) after we cancel the product $a_1a_2\ldots a_r$, a legal operation since $gcd(a_i,n) = 1$ for each $i = 1, 2, \ldots, r$. Thus, Euler proved the following generalization of Fermat's Little Theorem.

Theorem: Let $gcd(a,n) = 1$. Then $a^{\phi(n)} = 1$ (*mod n*). □

Observe that when n is a prime p, we have $\phi(n) = \phi(p) = p - 1$, in which case we get Fermat's Little Theorem, $a^{p-1} = 1$ (*mod p*) as a special case of the above more general theorem.

Example: If $gcd(k,10) = 1$, then $k^4 = 1$ (*mod* 10), implying that it ends with a 1. Thus, $3^4 = 81 = 1$ (*mod* 10).

PHI OF A PRODUCT OF *m* AND *n* WHEN gcd(*m,n*) > 1

Here is a theorem concerning $\phi(mn)$ when *m* and *n* are not necessarily relatively prime. We then cannot conclude that $\phi(mn) = \phi(m)\phi(n)$.

Theorem: $\phi(mn) = \dfrac{\phi(m)\phi(n)d}{\phi(d)}$, where $d = gcd(m,n)$.

Proof: Note that the prime divisors of *mn* are precisely the prime divisors of either *m* or *n* (or both). Let *M* be the product of the factors of the form $\left(1 - \dfrac{1}{p}\right)$ for each prime divisor of *m*, let *N* be the analogous product for *n*, and let *D* be the analogous product for *d*. It then follows, using formula (7.1) that $\dfrac{\phi(mn)}{mn}$ = the product of the factors of the form $\left(1 - \dfrac{1}{p}\right)$ for each prime divisor of *mn*. But this is precisely $\dfrac{MN}{D}$, since division by *D* removes one of each of the factors of the form $\left(1 - \dfrac{1}{p}\right)$ that appear twice in the numerator. Now, $M = \dfrac{\phi(m)}{m}$, $N = \dfrac{\phi(n)}{n}$, and $D = \dfrac{\phi(d)}{d}$. Then we have $\dfrac{\phi(mn)}{mn} = \dfrac{MN}{D} = \dfrac{\phi(m)\phi(n)d}{mn\phi(d)}$, from which we obtain $\phi(mn) = \dfrac{\phi(m)\phi(n)d}{\phi(d)}$.   □

Of course, when $gcd(m,n) = 1$, the above theorem reduces to $\phi(mn) = \phi(m)\phi(n)$, as one would expect, since ϕ is multiplicative.

THE ORDER OF *a* (*mod n*)

Given *a*, such that $1 \le a < n$ and $gcd(a,n) = 1$, we define the *order* of *a* (*mod n*) denoted $|a|$ (not to be confused with "absolute value") as the smallest positive exponent, *k*, such that $a^k = 1$ (*mod n*). Such a *k* must exist in light of Euler's theorem and must satisfy $k \le \phi(n)$. If $gcd(a,n) = d > 1$, the order of *a* does not make sense, since it is then impossible for $a^k = 1$ (*mod n*). To see this, note that $a^k = 1$ (*mod n*) is equivalent to the equation $a^k = 1 + jn$ for some *j*. But this is absurd since $d \mid a^k$, while *d* does not divide $1 + jn$.

Example: When $n = 7$, $|2| = 3$, since $2^3 = 1$ (*mod* 7) and 3 is the smallest exponent that does this. Note that $\phi(7) = 6$ and $|2| < \phi(7)$.

Example: When $n = 8$, $|3| = 2$, since $3^2 = 1$ (*mod* 8) and 2 is the smallest exponent that does this. Note that $\phi(8) = 4$ and, as in the previous example, $|3| < \phi(8)$.

Example: When $n = 7$, $|3| = 6$, since $3^6 = 1$ (*mod* 7) and 6 is the smallest exponent that does this. Note that $\phi(7) = 6$ and here $|3| = \phi(7)$, in contrast to the previous two examples where we obtained strict inequality.

Suppose $|a| = k$ (*mod* n), then we claim that $a^j = 1$ (*mod* n) if and only if $k \mid j$. The "if" part is easy. If $k \mid j$, then $j = km$. Then $a^j = a^{km} = (a^k)^m = 1^m = 1$ (*mod* n). For the "only if" part, assume that $a^j = 1$ (*mod* n) and using the Euclidean division equation, $j = qk + r$, where $0 \le r < k$. Then $a^j = a^{qk+r} = (a^k)^q a^r = a^r$ (*mod* n). Since $a^j = 1$ (*mod* n), we have $a^r = 1$ (*mod* n), implying that $r = 0$, since $r < k$ and k is the order of a. Then $j = qk$, implying that $k \mid j$. We have proved the following:

Theorem: If $|a| = k$ (*mod* n), then $a^j = 1$ (*mod* n) if and only if $k \mid j$. □

Corollary: If $|a| = k$ (*mod* n), then $k \mid \phi(n)$.

Proof: Since $a^{\phi(n)} = 1$ (*mod* n), it follows by the preceding theorem that $k \mid \phi(n)$. □

Corollary: If $|a| = k$ (*mod* p), where p is an odd prime, then $k \mid p - 1$.

Proof: This follows immediately by the previous corollary since $\phi(p) = p - 1$. □

Example: When $n = 7$, $|2| = 3$. Since $\phi(7) = 6$, we see that $|2| \mid \phi(7)$, as the corollary predicts.

Example: Prove that if $|a| = k$ (*mod* n) and $a^s = a^t$ (*mod* n), then $s = t$ (*mod* k). Assume that $s > t$. Then we cancel a^t in $a^s = a^t$ (*mod* n), yielding $a^{s-t} = 1$ (*mod* n), in which case the previous theorem permits us to conclude that $k \mid (s - t)$. Then $s = t$ (*mod* k).

PRIMITIVE ROOTS

If $|a|$ *mod* n is $\phi(n)$, then a is called a *primitive root* of n. Thus, 3 is a primitive root of 7 since $3^6 = 1$ (*mod* 7) and 6 is the smallest exponent that does this. On the other hand, 2 is not a primitive root of 7 since

$2^3 = 1 \ (mod \ 7)$ and $\phi(7) = 6$. Similarly, 3 is not a primitive root of 8 since $3^2 = 1 \ (mod \ 8)$ and $\phi(8) = 4$.

Theorem: If a is a primitive root of the odd prime p, then $a^{\frac{p-1}{2}} = -1 (mod \, p)$.

Proof: Let $a^{\frac{p-1}{2}} = x$. Then $x^2 = a^{p-1} = 1 \ (mod \ p)$, so $x^2 - 1 = 0 \ (mod \ p)$. Then $p \mid (x-1)(x+1)$ implying that either $p \mid (x-1)$ or $p \mid (x+1)$. Then either $x + 1 = 0 \ (mod \ p)$ or $x - 1 = 0 \ (mod \ p)$. This implies that either $x = -1 \ (mod \ p)$ or $x = 1 \ (mod \ p)$. The second equation is impossible since a is a primitive root. This leaves $x = -1 \ (mod \ p)$ or $a^{\frac{p-1}{2}} = -1 (mod \, p)$. $\square$

Corollary: If r and s are primitive roots of the odd prime p, then rs is not.

Proof: By the previous theorem, $r^{\frac{p-1}{2}} = -1 (mod \, p)$ and $s^{\frac{p-1}{2}} = -1 (mod \, p)$. It follows that $(rs)^{\frac{p-1}{2}} = 1 (mod \, p)$, in which case rs is not a primitive root. $\square$

While some numbers such as 15 have no primitive roots (see the exercises at the end of the chapter), it can be shown using group theory (a branch of mathematics beyond the scope of this text) that every odd prime, p, has a primitive root, a, where $2 \le a \le p - 2$. The first $p - 1$ powers of a (*mod p*) generate all the numbers 1, 2, ..., $p - 1$, in some order. That is, the set of numbers a, a^2, a^3, ..., a^{p-1} is a complete set of nonzero residues of p. Necessarily, $|a| = p - 1$. Of course $a \ne 1$ or $p - 1$, since $1^2 = (p - 1)^2 = 1 (mod \ p)$.

THE INDEX OF *m* (*mod p*) RELATIVE TO *a*

As a consequence of the preceding paragraph about the primitive root a for the odd prime p, given a number m such that $1 \le m < p$, there exists an exponent r such that $a^r = m \ (mod \ p)$. We call r the *index of m* (*mod p*) *relative to a*. In symbols, we write $ind_a m = r \ (mod \ p)$.

Example: $ind_6 8 = 3 \ (mod \ 13)$, since $6^3 = 8 \ (mod \ 13)$, which is $216 = 8$ (*mod* 13). By the way, this cumbersome calculation can be done as follows. Since $6^2 = 36 = 10 \ (mod \ 13)$, it follows that $6^3 = 6^2 \times 6 = 10 \times 6 = 60 = 8 \ (mod \ 13)$.

Example: $ind_6 10 = 2 \ (mod \ 13)$, since $6^2 = 10 \ (mod \ 13)$, which is $36 = 10$ (*mod* 13).

Here is a chart of the indices, *mod* 13, using the primitive root $a = 6$.

m	1	2	3	4	5	6	7	8	9	10	11	12
$ind_6 m$	12	5	8	10	9	1	7	3	4	2	11	6

Let us calculate $ind_6(2 \times 3)$. Now, $ind_6 2 = 5$ and $ind_6 3 = 8$, meaning that $6^5 = 2$ (*mod* 13) and $6^8 = 3$ (*mod* 13). It follows, upon multiplying these two equations, that $6^{13} = 6$ (*mod* 13). Since, by Fermat's Little Theorem, $6^{12} = 1$ (*mod* 13), this becomes $6^1 = 6$ (*mod* 13), showing that $ind_6(2 \times 3) = [ind_6 2 + ind_6 3]$ (*mod* 12), since $6^{12} = 1$ (*mod* 13).

Here is another example. To calculate $ind_6(3 \times 4)$, observe that $ind_6 3 = 8$ and $ind_6 4 = 10$, meaning that $6^8 = 3$ (*mod* 13) and $6^{10} = 4$ (*mod* 13). It follows, upon multiplying these two equations, that $6^{18} = 12$ (*mod* 13). Since, by Fermat's Little Theorem, $6^{12} = 1$ (*mod* 13), this becomes $6^6 = 12$ (*mod* 13), from which it follows that $ind_6 12 = 6$. Then $ind_6(3 \times 4) = [ind_6 3 + ind_6 4]$ (*mod* 12) $= 18$ (*mod* 12) $= 6$.

Observe that $ind_6 1 = 0$ (*mod* 12) and observe that $ind_6 8 = ind_6 2^3 = 3 \times ind_6 2 = 3 \times 5 = 3$ (*mod* 12), in agreement with the value of $ind_6 8$ in the above chart. All of these observations suggest the following parallels between indices and logarithms:

1. $ind_a mn = [ind_a m + ind_a n]$ (*mod* $p - 1$).
2. $ind_a m^n = [n \times ind_a m]$ (*mod* $p - 1$).
3. $ind_a 1 = 0$ (*mod* $p - 1$).
4. $ind_a a = 1$ (*mod* $p - 1$).

Example: Solve $x^2 = 10$ (*mod* 13). Using Law 2, $2 \times ind_6 x = ind_6 10$ (*mod* 12), or $2 \times ind_6 x = 2$ (*mod* 12). This equation becomes $ind_6 x = 1$ (*mod* 6), implying that $ind_6 x = 1$, in which case $x = 6$ or $ind_6 x = 7$, in which case $x = 7$. It is easily verified that $6^2 = 36 = 10$ (*mod* 13) and $7^2 = 49 = 10$ (*mod* 13).

Example: Solve $x^3 = 12$ (*mod* 13). Using Law 2, $3 \times ind_6 x = ind_6 12$ (*mod* 12), or $3 \times ind_6 x = 6$ (*mod* 12). This equation becomes $ind_6 x = 2$ (*mod* 4), implying that $ind_6 x = 2$, 6, or 10, yielding the solutions $x = 10$, 12, and 4, respectively.

Let a be a primitive root of the odd prime p. Then as we observed in the preceding section, the numbers a, a^2, a^3, ..., a^{p-1} constitute a complete set of nonzero residues of p. Furthermore, the order of a^r for any r between 1 and $p - 1$ divides $p - 1$, that is, $(|a^r|) | p - 1$. We shall determine the order of a^r as a function of r, that is, given r, we will determine $|a^r|$. Toward this end, let $gcd(r, p-1) = d$. Then $r = ud$ and $p - 1 = vd$, where $gcd(u,v) = 1$. (See section "Greatest Common Divisor" of Chapter 4 concerning the gcd.) We claim that $|a^r| = \dfrac{p-1}{d} = v$. We have $(a^r)^v = (a^{ud})^v = (a^{vd})^u = (a^{p-1})^u = 1$ (*mod p*). Furthermore, we claim that v is the smallest number that does this. To see this, let $|a^r| = k$. Then $a^{rk} = 1$ (*mod p*). It follows that $p - 1 \mid rk$. Rewrite this as $vd \mid udk$, which is equivalent to $v \mid uk$. Now, since $gcd(u,v) = 1$, this requires that $v \mid k$. Clearly, the smallest value of k that satisfies this is v. We have proven the following theorem.

Theorem: Let a be a primitive root of the odd prime p. Then $|a^r| = \dfrac{p-1}{d}$, where $d = gcd(r, p-1)$. ◻

Corollary: The number of primitive roots of the odd prime p is $\phi(p-1)$.

Proof: Since a^r is a primitive root of the odd prime p if and only if $|a^r| = p - 1$, we require that $d = gcd(r, p-1) = 1$, that is, r and $p-1$ must be relatively prime. Since $r \leq p-1$, the number of such r's is $\phi(p-1)$. ◻

Example: How many primitive roots does 7 have? Find them. By the above corollary, there are $\phi(6) = 2$ primitive roots. They are 3 and 5. We get 3 by trial and error. Then since 1 and 5 are relatively prime to 6, the two primitive roots are $3^1 = 3$ (*mod 7*) and $3^5 = 243 = 5$ (*mod 7*).

Example: How many primitive roots does 11 have? Find them. There are $\phi(10) = 4$ primitive roots. They are 2, 6, 7, and 8. We get 2 by trial and error. Then we observe that the integers 1, 3, 7, and 9 are relatively prime to 10. Then the four primitive roots of 11 are $2^1 = 2$ (*mod 11*), $2^3 = 8$ (*mod 11*), $2^7 = 128 = 7$ (*mod 11*), and $2^9 = 512 = 6$ (*mod 11*).

Now, suppose $d \mid p - 1$, where p is an odd prime with primitive root a. Then $p - 1 = dk$, in which case $|a^k| = d$. This can be seen as follows. Firstly, $(a^k)^d = a^{p-1} = 1$. Secondly, for any $r < d$, we have $kr < kd$, in which case $a^{kr} \neq 1$, since a is a primitive root. So $|a^k| = d$. Then given any divisor d of the odd prime p with primitive root a, there is at least one number of order d, namely, a^k, where $k = \dfrac{p-1}{d}$. This raises the question: how many numbers have order d?

Consider the d numbers a^k, a^{2k}, a^{3k}, ..., a^{dk}, no two of which are congruent *mod p*. (To see this, note that if $a^{ik} = a^{jk}$ *(mod p)* where $1 \le i < j \le d$, then we would have $1 = a^{(i-j)k}$ *(mod p)* where $(i-j)k < dk = p - 1$, contradicting the fact that a is a primitive root.)

Clearly, any number of this form, that is, a^{tk} where $1 \le t \le d$, satisfies $(a^{tk})^d = (a^t)^{dk} = (a^t)^{p-1} = 1$ *(mod p)*. This does not imply that $|a^{tk}| = d$. An exponent $r < d$ might do the job. In other words, perhaps $(a^{tk})^r = 1$ *(mod p)*. In fact, this will occur precisely when $gcd(t,d) > 1$, as can be seen as follows. Let $gcd(t,d) = s > 1$. Then $t = us$ and $d = vs$, where $gcd(u,v) = 1$. Then letting $r = v < d$, we have $(a^{tk})^r = (a^{usk})^v = (a^u)^{vsk} = (a^u)^{dk} = (a^u)^{p-1} = 1$ *(mod p)*. Now, according to a corollary to Lagrange's theorem (see Chapter 5), the equation $x^d - 1 = 0$ *(mod p)* has exactly d distinct roots *mod p*. Letting $x = a^k$, this implies that the $\phi(d)$ numbers less than, and relatively prime to d yield precisely those exponents t for which $(a^k)^t = 1$ *(mod p)*. Then letting $g(d)$ be the number of incongruent integers of order d, we have $g(d) \le \phi(d)$. It follows that $\sum_{d|p-1} g(d) \le \sum_{d|p-1} \phi(d)$. On the other hand, both sums must equal $p - 1$, the first because each of the $p - 1$ positive integer less than p must have an order that divides $p - 1$ and the second by Equation 7.2, which proves the following theorem.

Theorem: Let $d \mid p - 1$, where p is an odd prime. Then exactly $\phi(d)$ positive integers less than p have order d *(mod p)*.   □

Note that this theorem generalizes the previous corollary that stated that the number of primitive roots of the odd prime p is $\phi(p - 1)$.

Example: How many positive integers less than 7 have order 3 *(mod 7)*? Find them. By the above theorem, there are $\phi(3) = 2$ integers. They are 2 and 4. We know 3 is a primitive root. Then $(3^2)^3 = 3^6 = 1$ *(mod 7)*, implying that $3^2 = 2$ *(mod 7)* has order 3. Since 1 and 2 are relatively prime to 3, the integers $2^1 = 2$ and $2^2 = 4$ have order 3.

Example: How many positive integers less than 5 have order 2 *(mod 5)*? Find them. By the above theorem, there is $\phi(2) = 1$ integer, namely, 4, since $4^2 = 16 = 1$ *(mod 5)*.

Theorem: Let $gcd(a,p) = 1$, where p is an odd prime. Then the equation $x^k = a$ *(mod p)* is solvable if and only if $a^{\frac{p-1}{d}} = 1 \,(mod\,p)$, where $d = gcd(k, p - 1)$.

Proof: Assume that $x^k = a \ (mod \ p)$. Then raise both sides to the exponent $\frac{p-1}{d}$, yielding $x^{k\left(\frac{p-1}{d}\right)} = a^{\frac{p-1}{d}}(mod\,p)$. Since $d = gcd(k, \ p-1)$, it follows that $d \mid k$. Then let $y = x^{\frac{k}{d}}$. Then $x^{k\left(\frac{p-1}{d}\right)} = y^{p-1} = 1(mod\,p)$, by Fermat's Little Theorem. So $a^{\frac{p-1}{d}} = 1(mod\,p)$.

For the second part of the proof, we assume that $a^{\frac{p-1}{d}} = 1(mod\,p)$, and we must show that $x^k = a \ (mod \ p)$ is solvable. Letting r be a primitive root, this equation is solvable if and only if $k \times ind_r x = ind_r a \ (mod \ p-1)$. This is solvable if and only if $d \mid ind_r a$. Now, $a^{\frac{p-1}{d}} = 1(mod\,p)$ implies that $\frac{p-1}{d} \times ind_r a = 0(mod\,p-1)$, or $\frac{p-1}{d} \times ind_r a = m(p-1)$, which implies, upon canceling $p-1$, that $d \mid ind_r a$. The proof is over. $\qquad\square$

Corollary: Let p be an odd prime. Then $x^2 = -1 \ (mod \ p)$ has a solution if and only if $p = 1 \ (mod \ 4)$. In other words, -1 is a *quadratic residue* of p if and only if $p = 1 \ (mod \ 4)$.

Proof: By the theorem above, $x^2 = -1 \ (mod \ p)$ has a solution if and only if $(-1)^{\frac{p-1}{d}} = 1(mod\,p)$, where $d = gcd(2, p-1)$. Now, d is 2, since $p-1$ is even. So the condition $(-1)^{\frac{p-1}{d}} = 1(mod\,p)$ becomes $(-1)^{\frac{p-1}{2}} = 1(mod\,p)$. This requires that $\frac{p-1}{2}$ be even, that is, $\frac{p-1}{2} = 2k$, for some k, implying that $p = 4k+1$, that is, $p = 1 \ (mod \ 4)$. $\qquad\square$

Note that this corollary was presented as a theorem in Chapter 5.

TO BE OR NOT TO BE A QUADRATIC RESIDUE

Recall that a is a *quadratic residue mod p*, if $x^2 = a \ (mod \ p)$ has a solution. Otherwise, x is called a *quadratic nonresidue*. The following interesting corollary follows by the previous theorem upon letting $k = 2$, in which case $d = gcd(2, \ p-1) = 2$.

Corollary: Let $gcd(a,p) = 1$, where p is an odd prime. Then a is a quadratic residue if and only if $a^{\frac{p-1}{2}} = 1(mod\,p)$. $\qquad\square$

Corollary: Let $gcd(a,p) = 1$, where p is an odd prime. If a is a quadratic nonresidue, then $a^{\frac{p-1}{2}} = -1(mod\,p)$.

Proof: Let $y = a^{\frac{p-1}{2}}$. Then $y^2 = a^{p-1} = 1$, by Fermat's Little Theorem. Then $y^2 - 1 = 0 \ (mod \ p)$, implying that $y = 1 \ (mod \ p)$ or $y = -1 \ (mod \ p)$. Since a is a quadratic nonresidue, $y \neq 1 \ (mod \ p)$, implying that $y = -1$ $(mod \ p)$, that is, $a^{\frac{p-1}{2}} = -1 (mod \, p)$.   □

To summarize the above two corollaries, we have the following theorem.

Theorem: Let $gcd(a,p) = 1$, where p is an odd prime. Then a is a quadratic residue if $a^{\frac{p-1}{2}} = 1 (mod \, p)$ and a is a quadratic nonresidue if $a^{\frac{p-1}{2}} = -1 (mod \, p)$.

THE LEGENDRE SYMBOL

The preceding theorem motivated the French mathematician Legendre (1752–1833) to define the symbol (a/p) where $gcd(a,p) = 1$ and p is an odd prime, as follows:

$$(a/p) = \begin{cases} 1 & \text{if } a \text{ is a quadratic residue of } p \\ -1 & \text{if } a \text{ is a quadratic nonresidue of } p \end{cases}$$

(a/p) is called a *Legendre symbol*. Its use simplifies matters. The above remark, for example, can be stated as

$$(a/p) = a^{\frac{p-1}{2}} (mod \, p) \tag{7.4}$$

Here are a few properties of the Legendre symbols. The proofs are left as exercises. We assume that $gcd(a,p) = gcd(b,p) = 1$:

1. If $a = b \ (mod \ p)$, then $(a/p) = (b/p)$.
2. $(1/p) = 1$.
3. $(-1/p) = (-1)^{\frac{p-1}{2}}$.
4. $(a^2/p) = 1$.
5. $(ab/p) = (a/p)(b/p)$, that is, (a/p) is multiplicative.
6. $(ab^2/p) = (a/p)$.

Property 3 implies the following remark.

Remark: Let p be an odd prime. Then $(-1/p) = \begin{cases} 1 & \text{if } p = 1 \ (mod \ 4) \\ -1 & \text{if } p = 3 \ (mod \ 4) \end{cases}$

□

This remark restates the corollary in the previous section: "Let p be an odd prime. Then $x^2 = -1$ (*mod p*) has a solution if and only if $p = 1$ (*mod* 4). In other words, -1 is a *quadratic residue* of p if and only if $p = 1$ (*mod* 4)."

We shall use the above remark to show that there are infinitely many primes of the form $4k + 1$.

Theorem: There are infinitely many primes of the form $4k + 1$.

Proof: Using proof by contradiction, assume that there are only finitely many primes of the form $4k + 1$. Denoting them by $p_1, p_2, \ldots, p_r$, let $P = p_1 \times p_2 \times \cdots \times p_r$ and let $N = 4P^2 + 1$. Since N is odd, it has an odd prime factor, say, p. Then $p \mid 4P^2 + 1$, or in other words, $4P^2 = -1$ (mod p). Then $(-1/p) = 1$, implying that $p = 1$ (*mod* 4). But this is impossible, since p would then divide P. This would, in turn, imply that $p \mid 4P^2$, in which case p does not divide $4P^2 + 1$, a contradiction. □

Example: Use the properties of the Legendre symbols to determine whether $x^2 = 20$ (*mod* 31) is solvable. We have $(20/31) = (4/31)(5/31) = (5/31)$. Since $5 = 36$ (*mod* 31), we have $(5/31) = (36/31) = 1$. Alternatively, we can observe that $(5/31) = 5^{15}$ (*mod* 31). This is easy to evaluate since $5^3 = 125 = 1$ (*mod* 31), implying that $5^{15} = 1^3$ (mod 31) $= 1$. Thus, the equation is solvable.

Example: Let the odd prime p be the sum of two squares, that is, let $p = a^2 + b^2$. Show using the above remark that $p = 1$ (*mod* 4). Observe, first, that $1 \leq a < p$. (If $a = 0$, we would have the absurd statement $p = b^2$.) Since $gcd(a,p) = 1$, there exists an integer c such that $ac = 1$ (*mod p*). Now, $p = a^2 + b^2$ implies that $pc^2 = (ac)^2 + (bc)^2$, which becomes $(bc)^2 = -1$ (*mod p*). Then $(-1/p) = 1$, implying that $p = 1$ (*mod p*).

QUADRATIC RECIPROCITY

The following elegant theorem was discovered by Legendre who failed to prove it. Gauss supplied a proof at the age of 19 and went on to find six more proofs! We shall not prove it here.

LAW OF QUADRATIC RECIPROCITY

Theorem: Let p and q be distinct odd primes. Then $(p/q)(q/p) = (-1)^{\frac{p-1}{2} \cdot \frac{q-1}{2}}$. □

An immediate and very useful consequence is given by the following corollary.

Corollary: Let p and q be distinct odd primes. Then $(p/q) = (q/p)$ if at least one of p or q is congruent to 1 ($mod\ 4$). Otherwise, $(p/q) = -(q/p)$.

Proof: Without loss of generality, let $p = 1$ ($mod\ 4$). Then $p = 4k + 1$, in which case $\frac{p-1}{2} = 2k$, which is even. It then follows from the above theorem that $(p/q)(q/p) = 1$, implying that $(p/q) = (q/p)$. On the other hand, if $p = q = 3$ ($mod\ 4$), then $\frac{p-1}{2} \times \frac{q-1}{2}$ is odd. It then follows that $(p/q)(q/p) = -1$, implying that $(p/q) = -(q/p)$. □

Example: Evaluate $(3/13)$. Since $13 = 1$ ($mod\ 4$), we may invoke quadratic reciprocity. Then $(3/13) = (13/3) = (1/3)$ by Property 1 since $13 = 1$ ($mod\ 3$). Finally, by Property 2, $(1/3) = 1$, so $(3/13) = 1$, implying that 3 is a quadratic residue of 13.

WHEN DOES $x^2 = a$ (*mod n*) HAVE A SOLUTION?

When does $x^2 = a$ ($mod\ n$), where n is a composite number, have a solution? We begin our analysis when $n = p^m$ where p is an odd prime. Here is an important theorem toward this end.

Theorem: Let p be an odd prime and let $gcd(a,p) = 1$. Then the equation $x^2 = a$ ($mod\ p^m$) has a solution if and only if the equation $x^2 = a$ ($mod\ p$) has a solution.

Proof: As to the "only if" part of the theorem, the equation $x^2 = a$ ($mod\ p^m$) is equivalent to the statement $p^m \mid x^2 - a$, which implies that $p \mid x^2 - a$, or equivalently $x^2 = a$ ($mod\ p$). For the "if" part, we shall use induction on m. Since we are assuming that $x^2 = a$ ($mod\ p$) has a solution, the base case $m = 1$ is done. We assume that $x^2 = a$ ($mod\ p^k$) has a solution, x_0, and must prove that $x^2 = a$ ($mod\ p^{k+1}$) has a solution. Now, $x_0^2 = a$ ($mod\ p^k$) can be written $x_0^2 = a + cp^k$, for some c. Now, find an integer r such that $2x_0 r = -c \pmod{p}$. The integer $x_0 + rp^k$ solves the equation $x^2 = a \pmod{p^{k+1}}$. To show this, we have $(x_0 + rp^k)^2 = x_0^2 + 2x_0 rp^k + r^2 p^{2k} = a + cp^k + 2x_0 rp^k + r^2 p^{2k} = a + (c + 2x_0 r)p^k + r^2 p^{2k}$. Since $2x_0 r = -c \pmod{p}$, $p \mid c + 2x_0 r$, implying that $(c + 2x_0 r)p^k = 0 \pmod{p^{k+1}}$. Then $(x_0 + rp^k)^2 = a \pmod{p^{k+1}}$, as we claimed, and the proof is over. $\qquad\square$

Example: Show that 2 is a quadratic residue of 49. Since $3^2 = 2 \pmod{7}$, we see that 2 is a quadratic residue of 7. The previous theorem then shows that 2 is a quadratic residue of 49. (Note, by the way, that $10^2 = 100 = 2 \pmod{49}$.) More generally, it follows that 2 is a quadratic residue of 7^n, where $n \geq 1$.

Since the previous theorem deals with powers of an odd prime, it is natural to ask when is an odd number, a, a quadratic residue *mod* 2^n? Since $a = 1 \pmod{2}$, we see that every odd number is a quadratic residue *mod* 2. Since all odd number are congruent to either 1 or 3 *mod* 4, and since 1 is a quadratic residue of 4 while 3 is not, it follows that the odd number, a, is a quadratic residue *mod* 4 if and only if $a = 1 \pmod{4}$. Similarly, the only odd quadratic residue of 8 is 1, since $1^2 = 3^2 = 5^2 = 7^2 = 1 \pmod{8}$.

To complete the analysis of the odd quadratic residues of 2^n, we have the following:

Theorem: The odd integer a is a quadratic residue of 2^n, for $n \geq 3$, if and only if $a = 1 \pmod{8}$.

Proof: We use induction on n. The base case, $n = 3$, is obvious. So assume that for some $k > 3$, there is an x_0 such that $x_0^2 = a \pmod{2^k}$, or $x_0^2 = a + c2^k$, for some c. Now, find an integer r such that $x_0 r = -c \pmod{2}$. This can be done since x_0 is odd. The integer $x_0 + r2^{k-1}$ solves the equation $x^2 = a \pmod{2^{k+1}}$. To show this, $(x_0 + r2^{k-1})^2 = x_0^2 + x_0 r2^k + r^2 2^{2k-2} = a + c2^k + x_0 r2^k + r^2 2^{2k-2} = a + (c + x_0 r)2^k + r^2 2^{2k-2}$. Since $x_0 r = -c \pmod{2}$, it follows that $2 \mid c + 2x_0 r$, implying that $(c + x_0 r)2^k = 0 \pmod{2^{k+1}}$. Then $(x_0 + r2^{k-1})^2 = a \pmod{2^{k+1}}$ and the proof is over. $\qquad\square$

Example: Show that 17 is a quadratic residue of 128. Since $17 = 1 \pmod{8}$, and $128 = 2^7$, the previous theorem guarantees that 17 is a quadratic residue of 128.

The next theorem combines the last few results. The proof is left as an exercise.

Theorem: Let n have the prime decomposition $2^k p_1^{e_1} p_2^{e_2} \ldots p_r^{e_r}$ and let a be an integer such that $gcd(a,n) = 1$. Then a is a quadratic residue of n if and only if:

    1. a is a quadratic residue of $p_1, p_2, \ldots,$ and p_r.
    2. $a = 1 \ (mod \ 4)$ if $k = 2$.
    3. $a = 1 \ (mod \ 8)$ if $k \geq 3$.   □

Example: Show that 65 is a quadratic residue of 112. The prime decomposition of 112 is $2^4 7$. Since $65 = 1 \ (mod \ 8)$, and since $65 = 2$ (mod 7) and 2 is a quadratic residue of 7, it follows that 65 is a quadratic residue of 112.

EXERCISES

An asterisk (∗) indicates that the exercise can be developed into a research project.

1   Find $\phi(1000)$, $\phi(5040)$, $\phi(10^n)$, and $\phi(10!)$.

2   Verify the following statements for $n \leq 7$:

    Show that $\phi(3^n) = 2 \times 3^{n-1}$. Prove this for all n.
    If n is even, show that $\phi(2n) = 2 \times \phi(n)$.
    Let $n = 2^m$, where m is odd. Show that $\phi(n)$ is a square.
    If $3 \mid n$, show that $\phi(3n) = 3 \times \phi(n)$.
    If 3 does not divide n, show that $\phi(3n) = 2 \times \phi(n)$.
    If $m \mid n$, show that $\phi(m) \mid \phi(n)$.

3   Let p be a prime. Show that $p - 1 \mid \phi(p^k)$ for all $k \leq 7$ and $p \leq 101$.

4   Let p be a prime such that $p \mid n$. Show that $p - 1 \mid \phi(n)$ for all $n \leq 100$.

5   Find 100 values of k such that $10 \mid \phi(k)$.

6   Find 100 values of k such that $\phi(k) = k/3$.

7   Show that $\phi(n^2) = n\phi(n)$ for all $n \leq 100$.

8   ∗Show that $\phi(n^k) = n^{k-1}\phi(n)$ for all $n \leq 100$.

9   Find the last digit of 3^{345}. In other words, find the remainder when 3^{345} is divided by 10.

10 *For all values of n up to 50, show that the sum of the $\phi(n)$ numbers less than and relatively prime to n is 0 (*mod n*). Skip $n = 1$ and $n = 2$.

11 *Show that $\sum_{d|n} \mu(d)\phi(d) = (2-p_1)(2-p_2)...(2-p_r)$, where p_1, p_2, ..., p_r are the prime divisors of n.

12 *Show that if n is even, then $\sum_{d|n} \mu(d)\phi(d) = 0$. First, confirm this using a program for all $n \le 7$.

13 Show that $\sum_{d|n} \sigma(d) = n \sum_{d|n} \dfrac{\tau(d)}{d}$ for all $n \le 7$.

14 If f is an arithmetic function, show that $\sum_{d|n} f(d) = \sum_{d|n} f\left(\dfrac{n}{d}\right)$.

15 Find the order of 2 *mod p* and the order of 3 *mod p* for all primes between 5 and 101.

16 Find primitive roots for 11, 13, and 17.

17 *Show that 15 has no primitive roots by calculating the orders of 2, 4, 7, 8, 11, 13, and 14. Why does it suffice to calculate only these orders?

18 Show that 4 and 9 have primitive roots 3 and 2, respectively.

19 Given that 2 is a primitive root for 19, find all the primitive roots of 19.

20 How many primitive roots does 31 have? Do this in two ways.

21 *Let r be a primitive root for the odd prime p. Show that a is a quadratic residue *mod p* if and only if $ind_r a$ is even. Hint: if $ind_r a = 2m$ (*mod p*), then $a = r^{2m}$ (*mod p*).

22 *Show, using the previous exercise, that exactly half of the $p-1$ nonzero least residues of the odd prime p are quadratic residues.

23 Show that 2 is a primitive root for 19.

24 Given the prime modulus 17, find the number of least residues having each of the orders 1, 2, 4, 8, and 16. Then verify your answers by producing the residues having each of these orders.

25 If $a = b$ (*mod p*), show that $(a/p) = (b/p)$.

26 Show that $(1/p) = 1$.

27 Show that $(-1/p) = (-1)^{\frac{p-1}{2}}$.

28 Show that $(ab/p) = (a/p)(b/p)$, that is, show that (a/p) is multiplicative.

29 *Solve $5x^2 + 6x + 1 = 0$ (*mod* 23).

30 Show that 3 is a quadratic residue of 169.

31 Show that 5 is a quadratic residue of 1331.

32 Write a program to find $\phi(n)$. Then tabulate $\phi(n)$ for all $n \le 100$. *See the program at the end of the chapter.*

33 Given the positive integer m, write a program to find all $n \le 1000$ such that $\phi(n) = m$.

34 Use the program of the previous exercise to show that there is no $n \le 1000$ for which $\phi(n) = m$ when $m = 14$, 26, 34, 38, and 50. (In fact, there is no n at all for these values of m. There are infinitely many m's such that there is no n for which $\phi(n) = m$.) On the other hand, show that there are several values of n for which $\phi(n) = 12$.

35 The number 18 has an interesting property. $\phi(18) = 6$, since the numbers 1, 5, 7, 11, 13, and 17 are relatively prime to 18. All of these numbers (except 1) are prime. Find all numbers with this property. Hint: it is known that no number larger than 30 has this property.

36 Given the prime p and the positive integer a such that $gcd(a,p) = 1$, write a program to find $|a|$ (*mod* p).

The Euler Phi Function from Definition

```
<!DOCTYPE HTML>
<html>
<meta charset="UTF-8">
<head>
<title>PHI function</title>
<script>
 var placeref;
function find(){
 placeref = document.getElementById("place");
 placeref.innerHTML = "";
 n = parseInt(document.f.num.value);
```

```
 ans = phi(n);
 placeref.innerHTML = "The Euler phi function for
"+String(n)+" is "+String(ans);
 return false;
}
function phi(n) {
  var ans = 0;
  for (var i=1;i<n;i++){
    if (1 == gcd(i,n)){
      ans++;
    }
  }
  return ans;
}

function gcd(a,b){
  var holder;
  var r;
  if (b < a) {
  //switch

   holder = a;
   a = b;
   b = holder;
   }

  r = b % a;
  if (r==0){

    return a;
  }

  else {
    return gcd(r,a);

  }
}


</script>
</head>
<body>
```

```
Enter number <br/>
<form name="f" onsubmit="return find();">
 <input type="number" name="num" value=""/>  

  <input type="submit" value="Enter"/>
</form>
<p id="place">
  Messages will go here.
</p>
</body>
</html>
```

The Euler Phi Function from Formula

```
<!DOCTYPE HTML>
<html>
<meta charset="UTF-8">
<head>
<title>PHI from formula</title>
<script type="text/javascript"
src="primeDecomposition.js"> </script>
<script>

function phiFF(){
  var table;
  placeref = document.getElementById("place");
  n = parseInt(document.f.num.value);
  table = buildPrimeDecomposition(n);
  t = n;
  for (var i=0;i<table.length;i++){
    t = t * (1-(1/table[i][0]));
  }
t = Math.floor(t);
//correction for inaccuracies from floating pt
//arithmetic
  messages = "The value of PHI for "+ String(n)+" is "+
String(t);
  placeref.innerHTML = messages;
  return false;
}
</script>
</head>
<body>
```

```
Enter number to see its phi value:
<form name="f" onSubmit="return phiFF();">
 <input type="number" name="num" value=""/>
 <input type="submit" value="Enter integer"/>
</form>
<div id="place">
Answer will go here.
</div>
</body>
</html>
```

Numbers Whose Euler Phi Function Values
Are a Specified Number

```
<!DOCTYPE HTML>
<html>
<meta charset="UTF-8">
<head>
<title>PHI inverse</title>
<script>
 var placeref;
 var limit = 1000;

function find(){
 var m;
 placeref = document.getElementById("place");
 m = parseInt(document.f.num.value);
 message = "";
 placeref.innerHTML = message;

  for (var n = 1;n<=limit;n++){
   if (m==phi(n)){
    message+="<br/>"+String(n);

   }
  }
  if (message.length<1){
    message = "No values <= "+String(limit)+" have
phi equal to "+String(m);
  }
  else {
```

```
    message = "Values <= "+String(limit)+" with phi
equal to "+String(m)+"<br/>"+ message;
  }
  placeref.innerHTML = message;

 return false;
}
function phi(n) {
  var ans = 0;
  for (var i=1;i<n;i++){
    if (1 == gcd(i,n)){
      ans++;
    }
  }
  return ans;
}

function gcd(a,b){
  var holder;
  var r;
  if (b < a) {
  //switch

   holder = a;
   a = b;
   b = holder;
   }

  r = b % a;
  if (r==0){

     return a;
  }

  else {
    return gcd(r,a);

  }
}

</script>
</head>
<body>
```

```
Enter number <br/>
<form name="f" onsubmit="return find();">
 <input type="number" name="num" value=""/>  

  <input type="submit" value="Enter"/>
</form>
<p id="place">
  Messages will go here.
</p>
</body>
</html>
```

8

SUMS AND PARTITIONS

We will learn several interesting ways to express a given integer as a sum of special kinds of integers. These sums are called "partitions."

AN nTH POWER IS THE SUM OF TWO SQUARES

The following theorem shows how to find fourth powers that are the sums of two squares.

Theorem: There exist infinitely many solutions to the Diophantine equation $a^2 + b^2 = w^4$.

Proof: Let $u = 2st$, $v = s^2 - t^2$, and $w = s^2 + t^2$ be a Pythagorean triple. Then $a = 2uv$, $b = u^2 - v^2$, and $c = u^2 + v^2$ is also a Pythagorean triple. Then $a^2 + b^2 = c^2 = (u^2 + v^2)^2 = (w^2)^2 = w^4$. $\qquad\square$

The next theorem generalizes the last one. It shows that one can find two squares whose sum is an nth power for any n. We need the fact that the *length* of a complex number $w = a + ib$ (where $i = \sqrt{-1}$) is $\sqrt{a^2 + b^2}$ and is denoted $\|a + ib\|$. This is often written $\|w\|^2 = \|a + ib\|^2 = a^2 + b^2$.

Elementary Number Theory with Programming, First Edition. Marty Lewinter and Jeanine Meyer.
© 2016 John Wiley & Sons, Inc. Published 2016 by John Wiley & Sons, Inc.

So, for example, $\|3 + 4i\|^2 = 3^2 + 4^2 = 25$. It is shown in a branch of mathematics called complex analysis that if w is a complex number, then $\|w^n\| = \|w\|^n$. When $w = a + ib$ and $n = 2$, for example, we get $\|w^2\| = \|(a + ib)^2\| = \|a^2 - b^2 + 2abi\| = \sqrt{(a^2 - b^2)^2 + 4a^2 b^2} = \sqrt{a^4 - 2a^2 b^2 + b^4 + 4a^2 b^2} = \sqrt{a^4 + 2a^2 b^2 + b^4} = a^2 + b^2 = \|w\|^2$.

Theorem: Given the positive integers n and z such that z is the sum of two squares, there exist integers a and b such that $a^2 + b^2 = z^n$.

Proof: Let $z = r^2 + s^2$. Then expand the complex number $(r + is)^n$ yielding the complex number $a + ib$. Then $a^2 + b^2 = \|a + ib\|^2 = \|(r + is)^n\|^2 = \|r + is\|^{2n} = (\|r + is\|^2)^n = (r^2 + s^2)^n = z^n$. □

Example: Write 5^3 as the sum of two squares. Since $5 = 2^2 + 1^2$, we have $r = 2$ and $s = 1$. Now, $(2 + i)^3 = 2^3 + 3 \times 2^2 \times i + 3 \times 2 \times i^2 + i^3 = 8 + 12i - 6 - i = 2 + 11i$. Then $a = 2$ and $b = 11$. Indeed, $5^3 = 2^2 + 11^2$, or $125 = 4 + 121$.

Example: Write 17^3 as the sum of two squares. Since $17 = 4^2 + 1^2$, we have $r = 4$ and $s = 1$. Now, $(4 + i)^3 = 4^3 + 3 \times 4^2 \times i + 3 \times 4 \times i^2 + i^3 = 64 + 48i - 12 - i = 52 + 47i$. Then $a = 52$ and $b = 47$. Indeed, $17^3 = 52^2 + 47^2$, or $4913 = 2704 + 2209$.

SOLUTIONS TO THE DIOPHANTINE EQUATION $a^2 + b^2 + c^2 = d^2$

Here is a way of generating infinitely many solutions to the Diophantine equation $a^2 + b^2 + c^2 = d^2$:

$$a = u^2 + v^2 - w^2$$
$$b = 2uw$$
$$c = 2vw$$
$$d = u^2 + v^2 + w^2$$

For example, when $u = 1$, $v = 2$, and $w = 2$, we get $1^2 + 4^2 + 8^2 = 9^2$. More trivially, when $u = v = w = 1$, we get $1^2 + 2^2 + 2^2 = 3^2$.

Here is a way of generating infinitely many solutions to the Diophantine equation $a^2 + b^2 + c^2 = d^2$ such that $d = c + 1$. This becomes $a^2 + b^2 + c^2 = (c + 1)^2$, or $a^2 + b^2 = 2c + 1$. It follows that if a and b have opposite parity, we can let $c = \dfrac{a^2 + b^2 - 1}{2}$. When we let $a = 1$ and $b = 2$, for example, we get $c = 2$ and $d = c + 1 = 3$, yielding $1^2 + 2^2 + 2^2 = 3^2$, as we obtained earlier. When $a = 1$ and $b = 4$, we get $c = 8$ and $d = 9$, again in agreement with earlier results. Letting $a = 1$ and $b = t$, we have $c = \dfrac{t^2}{2}$, and $d = \dfrac{t^2}{2} + 1$, a more general result.

ROW SUMS OF A TRIANGULAR ARRAY OF CONSECUTIVE ODD NUMBERS

Consider the first few rows of an infinite triangular array of consecutive odd numbers:

$$
\begin{array}{ccccccc}
 & & & \mathbf{1} & & & \\
 & & 3 & & 5 & & \\
 & 7 & & \mathbf{9} & & 11 & \\
 13 & & 15 & & 17 & & 19 \\
 21 & 23 & & \mathbf{25} & & 27 & 29 \\
 31 & 33 & 35 & & 37 & 39 & 41 \\
 43 & 45 & 47 & \mathbf{49} & 51 & 53 & 55 \\
\end{array}
$$

   The row sums are 1, 8, 27, 64, 125, and 216. These numbers are the first six cubes, leading one to conjecture that the sum of the entries of the nth row is n^3. To see that this is in fact the case, observe that the average value of the nth row is n^2 and the nth row obviously has n entries. (The average actually appears (in bold font) as the middle entry in the odd rows.) Now, the average of a set of n numbers is their sum divided by n. Then the sum of the entries in the nth row is n times their average, that is, $n \times n^2 = n^3$. This was observed by Nicomachus (circa 100 A.D.).

PARTITIONS

A *partition* of the positive integer n is a representation of n as a sum of one or more positive integers. The seven partitions of 5 are $5, 4 + 1, 3 + 2$, $3 + 1 + 1, 2 + 2 + 1, 2 + 1 + 1 + 1$, and $1 + 1 + 1 + 1 + 1$.

The terms in a partition are *summands*. Unless otherwise stated, the order of the summands does not matter. $3 + 2$ and $2 + 3$ constitute the same partition of 5. If order matters, we have *ordered partitions* of n. Recall from Chapter 2 that the number of ordered partitions of n consisting of 1's and 2's $= u_{n+1}$, is the $(n + 1)$-st Fibonacci number.

It will be useful to represent a partition $a_1 + a_2 + \cdots + a_r$ of n consisting of r summands by a matrix of dots having r rows such that the first row has a_1 dots, the second row has a_2 dots, ..., and the rth row has a_r dots. See Figure 8.1 for the "dot matrix" representation of the partition $3 + 3 + 2 + 1$ of 9. Notice that it has four rows, since the given partition has four summands. Note, furthermore, that it has three columns since the maximum summand of the partition is 3.

We define the *transpose* of a matrix of dots in the same way that this is done for matrices with numerical entries. The rows of the transpose are the respective columns of the original matrix. The matrix of Figure 8.2 displays the transpose of the matrix of Figure 8.1. Notice that its rows represent another partition of 9, namely, $4 + 3 + 2$.

Observe that if a matrix has r rows and k columns, then the transpose has k rows and r columns. Note also that the matrix representation of a partition of n with r summands such that the maximum value of the summands is k has r rows and k columns, implying that the partition of n represented by its transpose has k summands and the maximum

FIGURE 8.1   A matrix representation of the partition $3 + 3 + 2 + 1$ of 9.

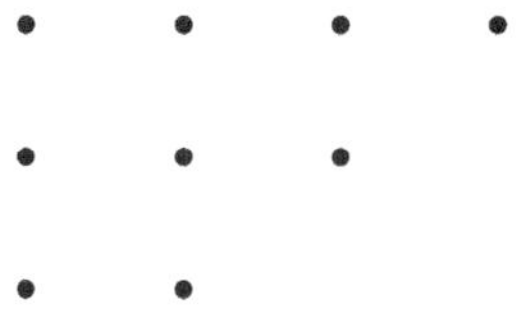

FIGURE 8.2   The transpose of the matrix of Figure 8.1.

value of the summands is r. This observation led Euler to state the following theorem.

Theorem: Given the positive integers n and k such that $k \leq n$, the number of partitions of n consisting of k or fewer summands equals the number of partitions of n such that no summand exceeds k.

Proof: There is a one-to-one correspondence between the two kinds of partitions described in the theorem. That is, to each partition of n consisting of k or fewer summands, we can find a unique partition of n such that no summand exceeds k and vice versa. This is done by taking the transpose of any matrix representing a partition in one of the groups. The transpose will represent exactly one corresponding partition in the other group.   □

The chart below demonstrates this correspondence for $n = 7$ and $k = 3$.

3 or fewer summands	Each summand ≤ 3
7	$1 + 1 + 1 + 1 + 1 + 1 + 1$
$6 + 1$	$2 + 1 + 1 + 1 + 1 + 1$
$5 + 2$	$2 + 2 + 1 + 1 + 1$
$5 + 1 + 1$	$3 + 1 + 1 + 1 + 1$
$4 + 3$	$2 + 2 + 2 + 1$
$4 + 2 + 1$	$3 + 2 + 1 + 1$
$3 + 3 + 1$	$3 + 2 + 2$
$3 + 2 + 2$	$3 + 3 + 1$

While counting the partitions of n is very difficult and is beyond the scope of this text, it is relatively easy to count the *ordered* partitions of n as our next theorem shows.

Theorem: There are 2^{n-1} ordered partitions of n.

Proof: Given n, list n 1's with spaces between them. An ordered partition of n may be determined by:

   a. Placing a vertical bar in some or all of the $n - 1$ spaces

   b. Adding the 1's between successive vertical bars

   c. Replacing the vertical bars by plus signs

Thus, the ordered partition $3 + 2 + 2 + 1$ of 8, for example, can be achieved as follows:

a. $1\ 1\ 1\ |\ 1\ 1\ |\ 1\ 1\ |\ 1$
b. $3\ |\ 2\ |\ 2\ |\ 1$
c. $3 + 2 + 2 + 1$

It follows that there are $n - 1$ "yes/no" choices to be made, yielding 2^{n-1} ways to insert the vertical bars, and the theorem is proven. □

A partition of n is called *odd* if each summand is odd. Thus, $5 + 3 + 3 + 1 + 1$ is an odd partition of 13, while $10 + 3$ is not. A partition of n is called *distinct* if each summand is distinct. Thus, $5 + 3 + 2 + 1$ is a distinct partition of 11, while $5 + 3 + 3$ is not, since the 3 is repeated. Of course, a partition can be both odd and distinct such as the partition $7 + 3 + 1$ of 11. It would seem that these two kinds of partitions are not related, but Euler showed that they are, as our next theorem asserts. Its proof, however, requires a lemma.

Lemma: If $|a| < 1$, then the infinite series $1 + a + a^2 + a^3 + \cdots = \dfrac{1}{1-a}$.

Proof: Recall that $1 + a + a^2 + a^3 + \cdots + a^{n-1} = \dfrac{a^n - 1}{a - 1}$. Now, when n goes to infinity, a^n goes to zero since $|a| < 1$, yielding the claim of the lemma. □

Theorem: The number of odd partitions of n equals the number of distinct partitions of n.

Proof: Let $Q(x) = (1 + x)(1 + x^2)(1 + x^3)(1 + x^4)\ldots$ This is an infinite product—something of great interest to Euler. Multiplying, we get $Q(x) = 1 + a_1 x + a_2 x^2 + a_3 x^3 + a_4 x^4 + \cdots$, where the coefficients have an interesting property. Consider a_4, for example. It equals the number of *distinct* partitions of 4, namely, 2, since x^4 is obtained by multiplying $1 \times x^4 = x^4$ and $x^1 \times x^3 = x^{1+3} = x^4$, representing the two distinct partitions, 4 and $1 + 3$, of 4. Note that there is no other way, such as $x^2 \times x^2$, to obtain x^4 since $Q(x)$ contains only one x^2. This corresponds to the fact that $2 + 2$ is not a distinct partition of 4.

Define $R(x) = \dfrac{1}{(1-x)(1-x^3)(1-x^5)\ldots} = \left(\dfrac{1}{1-x}\right)\left(\dfrac{1}{1-x^3}\right)\left(\dfrac{1}{1-x^5}\right)\cdots$

Applying the preceding lemma to each factor, we get

$$R(x) = \left(1 + x^1 + x^2 + x^3 + \cdots\right) \times \left(1 + x^3 + x^6 + x^9 + \cdots\right)$$
$$\times \left(1 + x^5 + x^{10} + x^{15} + \cdots\right)\ldots$$
$$= \left(1 + x^1 + x^{1+1} + x^{1+1+1} + \cdots\right) \times \left(1 + x^3 + x^{3+3} + x^{3+3+3} + \cdots\right)$$
$$\times \left(1 + x^5 + x^{5+5} + x^{5+5+5} + \cdots\right)\ldots$$
$$= 1 + b_1 x + b_2 x^2 + b_3 x^3 + b_4 x^4 + \cdots,$$

where the coefficient b_n counts the number of *odd* partitions of n, since the term x^n results by adding powers of terms in the parentheses of the second expression for $R(x)$. For example, x^{10} is obtained by multiplying $x^{1+1} \times x^3 \times x^5$, which corresponds to the odd partition $1 + 1 + 3 + 5$ for 10. The proof will be complete when we show that $Q(x) = R(x)$. Let

$$P(x) = (1-x)\left(1-x^2\right)\left(1-x^3\right)\left(1-x^4\right)\left(1-x^5\right)\ldots$$

Then

$$P(x)Q(x) = \left(1-x^2\right)\left(1-x^4\right)\left(1-x^6\right)\left(1-x^8\right)\ldots$$

Then $\dfrac{1}{Q(x)} = \dfrac{P(x)}{P(x)Q(x)} = (1-x)\left(1-x^3\right)\left(1-x^5\right)\left(1-x^7\right)\ldots = \dfrac{1}{R(x)},$

from which we get $Q(x) = R(x)$, and the proof is over.   □

Let us call a partition of n such that each summand is a distinct power of 2 (including 1 that is 2^0) a *distinct binary partition*. The first 10 positive integers have the distinct binary partitions displayed in the chart below. Note that if the binary partitions are not required to be distinct, we could include the partitions $2 + 2$, $2 + 1 + 1$, and $1 + 1 + 1 + 1$ for 4, for example.

We ask two questions:

1. Does every positive integer have a distinct binary partition?
2. Does any positive integer have more than one distinct binary partition?

n	Binary partition
1	1
2	2
3	$1 + 2$
4	4
5	$4 + 1$
6	$4 + 2$
7	$4 + 2 + 1$
8	8
9	$8 + 1$
10	$8 + 2$

The above chart leads us to suspect that the answers to questions 1 and 2 are *yes* and *no*, respectively. The next theorem shows that this is correct.

Theorem: Every positive integer has exactly one distinct binary partition.

Proof: Let $P(x) = (1 + x^1)(1 + x^2)(1 + x^4)(1 + x^8)(1 + x^{16})\ldots$ Multiplying, we get

$$P(x) = 1 + a_1 x + a_2 x^2 + a_3 x^3 + a_4 x^4 + \cdots, \qquad (8.1)$$

where the coefficient a_n of x^n equals the number of distinct binary partitions of n. Consider a_4, for example, it equals the number of distinct partitions of 4, namely, 1, since x^4 is obtained by multiplying $1 \times x^4 = x^4$. There is no other way to obtain x^4 in the infinite product P (x). The proof will be complete when we show that $a_n = 1$ for each $n = 1\ 2,\ 3,\ \ldots$.

Now, $\dfrac{P(x)}{1+x} = \left(1 + x^2\right)\left(1 + x^4\right)\left(1 + x^8\right)\left(1 + x^{16}\right)\ldots$. The product can be obtained by substituting an x^2 for each x that appears in the product for $P(x)$, that is, $\dfrac{P(x)}{1+x} = \left(1 + x^2\right)\left(1 + x^4\right)\left(1 + x^8\right)\left(1 + x^{16}\right)\ldots = P\left(x^2\right)$.

Now, from the relation (8.1), we get, upon replacing each x by x^2, $P\left(x^2\right) = 1 + a_1 x^2 + a_2 x^4 + a_3 x^6 + a_4 x^8 + \cdots$ so $\dfrac{P(x)}{1+x} = 1 + a_1 x^2 + a_2 x^4 + a_3 x^6 + a_4 x^8 + \cdots$, or, after transposing the $1 + x$, $P(x) = (1 + x)\left(1 + a_1 x^2 + a_2 x^4 + a_3 x^6 + a_4 x^8 + \cdots\right) = 1 + x + a_1 x^2 + a_1 x^3 + a_2 x^4 + a_2 x^5 + a_3 x^6 + a_3 x^7 + \cdots$. Now, we compare the coefficients in (8.1) with the coefficients in this expression for $P(x)$, yielding $a_1 = 1$, $a_2 = a_1 = 1$, a_3 $= a_1 = 1$, $a_4 = a_2 = 1$, ... We find indeed that $a_n = 1$ for each $n = 1$ 2, 3, ... $\square$

This theorem enables us to represent each nonnegative integer, n, by a unique *binary number* by putting a 1 in the kth position whenever the unique binary partition for n contains 2^{k-1}. The binary form of 13, for example, is 1101, since $13 = 1 + 4 + 8$, which can be written as 2^0 $+ 2^2 + 2^3$. Thus, binary numbers are expressed in "base 2" just as ordinary integers are said to be in "base 10" notation.

Binary numbers are essential to computer science, where 1 represents an electronic signal and 0 represents the absence of a signal. Put another way, 1 represents "circuit on" and 0 represents "circuit off." Thus, the basic arithmetic operations are performed electronically by circuit boxes that are wired to process incoming signals and emit outgoing signals in ways that correspond to arithmetic operations. Word processing is handled by assigning an integer (in binary form) to each letter and symbol (such as "?" and "%").

As suggested, computer scientists view binary partitions in a very special way. The idea is that integers are values that exist independent of representation. The way that we typically think of a number is in its decimal representation, but the number is independent of that. Remember back to the lessons on place value at school. The number 12 is the sum of 10 to the first power followed by 2. More accurately, it is 1 times 10 to the 1st power followed by 2 times 10 to the 0th power, which is defined to be 1. The number ten is written 10. Decimal made use of base 10 and other systems were possible. Because it is easier to build circuitry with just on and off states, base two became the choice for computers. Some early computers actually did use base 10, but this approach was soon discarded.

Joke (I have the T-shirt): There are 10 types of people in the world—those who understand binary and those who don't.

Because binary representation is how numbers are represented in computers, JavaScript provides a quick way to produce the binary representation: the integer method `toString`. This doesn't quite work for negative numbers so I use absolute value to correct any pesky user who tries to enter a negative number. You can research representation of negative numbers in JavaScript to see what you need to so.

Okay, my coauthor probably wanted me and you to do actual work to produce the unique binary partition. To this end, I wrote another program that you can examine at the end of the chapter. The approach is to determine what the largest power p of 2 bigger or equal to the number n. If 2^p is equal to the number, the program is essentially done. The representation is the character 1 followed by a string of p of the character 0. If that is not true, then my code decrements p, creates a string starting with a 1 and followed by a string of p (the new p) 0s, and continues with a *while* loop. The purpose of the *while* loop is to check if one of the 0s in the string needs to be changed to a 1. The code in the loop systematically checks for powers of 2. The value of n is changed (decreased) whenever it is shown that it is at least as big as a power of 2. The loop also makes use of a variable q that points to the position in the representation string that needs to be changed. The variable p decreases and the variable q increases.

WHEN IS A NUMBER THE SUM OF TWO SQUARES?

Here is a theorem of fundamental importance.

Theorem: Let p be a prime such that $p = 1$ (*mod* 4). Then p can be written as the sum of two squares.

Proof: Since $p = 1$ (*mod* 4), it follows that $(-1/p) = 1$. Then there exists an integer, m, such that $m^2 = -1$ (*mod p*), in which case $p \mid 1 + m^2$. Now, let S be the set of all so-called Gaussian integers, that is, complex numbers of the form $a + bi$, where a and b are integers. Divisibility is defined as follows. Given z, $w \in S$, then $z \mid w$ if there exists $u \in S$ such that $w = uz$. The fact that $p \mid 1 + m^2$ implies that $p \mid (1 + mi)(1 - mi)$ in S.

Now, if p is prime in S, it would follow that either $p \mid 1 + mi$ or $p \mid 1 - mi$. The statement $p \mid 1 + mi$ is equivalent to the equation $1 + mi = \alpha p$, where $\alpha \in S$. Taking the conjugate of both sides of this equation yields $1 - mi = \bar{\alpha} p$, implying that $p \mid 1 - mi$. A similar argument shows that if $p \mid 1 - mi$, then $p \mid 1 + mi$. It would then follow that $p \mid (1 + mi) + (1 - mi)$, or $p \mid 2$, which is absurd. Then p is composite in S, in which case we have $p = (a + bi)(c + di)$. Now, square the modulus of both sides of this equation yielding $p^2 = (a^2 + b^2)(c^2 + d^2)$. Then p^2 is the sum of two squares. $\square$

Recall, by the way, that a prime, p, satisfying $p = 3$ (*mod p*) cannot be written as the sum of two squares. Hence, we can say that an odd prime, p, can be written as the sum of two squares if and only if $p = 1$ (*mod* 4). Many composite numbers can be written as the sum of two squares, some in several ways. For example, $50 = 49 + 1 = 25 + 25$, and $125 = 100 + 25 = 121 + 4$. The next theorem is, therefore, somewhat surprising.

Theorem: Let p be a prime such that $p = 1$ (*mod* 4). Then p can be written as the sum of two squares in only one way except for the order of the squares.

Proof: Assume that $p = a^2 + b^2 = c^2 + d^2$. Then $c^2 = -d^2$ (*mod p*). Then we obtain

$$a^2d^2 - b^2c^2 = a^2d^2 - b^2(-d^2) = a^2d^2 + b^2d^2$$
$$= (a^2 + b^2)d^2 = pd^2 = 0(mod\,p)$$

or

$$a^2d^2 - b^2c^2 = 0(mod\,p),$$

implying that either

  a. $ad - bc = 0$ (*mod p*) or
  b. $ad + bc = 0$ (*mod p*).

Since a^2, b^2, c^2, and d^2 are strictly less than p, it follows that each of a, b, c, and d is strictly less than $\sqrt{p}$, implying that ad and bc are strictly less than p. Then cases (a) and (b) become

  a. $ad - bc = 0$ or
  b. $ad + bc = p$.

Since $p^2 = (a^2 + b^2)(c^2 + d^2) = (ad + bc)^2 + (ac - bd)^2$, if case (b) prevails, then we would have $p^2 = p^2 + (ac - bd)^2$, implying that $ac = bd$. If case (a) prevails, we have $ad = bc$. We claim that either of these possibilities ($ac = bd$ or $ad = bc$) implies the truth of the theorem. If $ad = bc$, then $a \mid bc$. Now, the equation $p = a^2 + b^2$ implies that $gcd(a,b) = 1$. Then $a \mid bc$ implies that $a \mid c$, or equivalently, $c = ka$. Then $ad = bc$ becomes

$ad = bka$, which yields $d = bk$. Then we obtain $p = c^2 + d^2 = (a^2 + b^2)k^2 = pk^2$, implying that $k = 1$. Then $c = a$ and $d = b$.

On the other hand, if $ac = bd$, then $c \mid bd$. Now, the equation $p = c^2 + d^2$ implies that $gcd(c,d) = 1$. Then $c \mid bd$ implies that $c \mid b$, or equivalently, $b = kc$. Then $ac = bd$ becomes $ac = kcd$, which yields $a = kd$. Then we obtain $p = a^2 + b^2 = (c^2 + d^2)k^2 = pk^2$, implying that $k = 1$. Then $a = d$ and $b = c$, completing the proof of the theorem. $\qquad\square$

Recall from Chapter 2 that if x and y can be written as the sum of two squares, then their product, xy, can also be written as the sum of two squares. This can be generalized to finite products of more than two factors. For example, if x, y, and z can be written as sums of two squares, then their product, xyz, can be written $(xy)z$. Now, each of xy and z can be written as sums of two squares, implying that xyz can also be written as the sum of two squares. Our next theorem states precisely when an integer is the sum of two squares.

Theorem: Let n be a positive integer such that $n = k^2m$, where m is square-free. Then n is the sum of two squares if and only if the prime decomposition of m contains no primes p such that $p = 3$ (*mod* 4).

Proof: For the "if" part of the proof, note that if m is the product of primes that are sums of two squares, then m is also the sum of two squares, that is, $m = a^2 + b^2$. Then we have $n = k^2m = k^2(a^2 + b^2) = (ka)^2 + (kb)^2$.

For the "only if" part, let $n = a^2 + b^2 = k^2m$, where m is square-free. If $m = 1$, we are done. So assume that $m > 1$. Let $gcd(a,b) = d$. Then $a = dr$ and $b = dt$, where $gcd(r,t) = 1$.

Then $d^2(r^2 + t^2) = k^2m$, implying that $d^2 \mid k^2$, since m is square-free. Then $d \mid k$, or equivalently, $k = ud$. Then $d^2(r^2 + t^2) = k^2m$ becomes $r^2 + t^2 = u^2m$. Now, let p be an odd prime divisor of m. It follows from the last equation that $r^2 + t^2 = 0$ (*mod* p). Since $gcd(r,t) = 1$, at least one of r and t, say, r, is relatively prime to p. Then for some r', we have $rr' = 1$ (*mod* p). Then multiply both sides of $r^2 + t^2 = 0$ (*mod* p) by $(r')^2$, yielding $1 + (tr')^2 = 0$ (*mod* p), or $(tr')^2 = -1$ (*mod* p), showing that $(-1/p) = 1$, in which case $p = 1$ (*mod* p), and the proof is over. $\qquad\square$

Corollary: The positive integer n is the sum of two squares if and only if in the prime decomposition of n, the exponent of each prime p such that $p = 3$ (*mod* 4) is even. $\qquad\square$

Example: Show that 3430 is not the sum of two squares. The prime decomposition of 3430 is $2 \times 5 \times 7^3$. Since $7 = 3 \ (mod\ 4)$, the claim follows from the above corollary.

Example: Show in two different ways that 2^n is the sum of two squares for each $n \geq 1$. The claim is a consequence of the above corollary. Alternatively, if $n = 2k$, we have $2^n = (2^k)^2 + 0^2$, while if $n = 2k + 1$, we have $2^n = 2(2^k)^2 = (2^k)^2 + (2^k)^2$.

Example: Show in two different ways that if $n = 3$ or $6 \ (mod\ 9)$, then n is not the sum of two squares. Since the exponent of 3 in the prime decomposition of n is 1, the claim is a consequence of the above corollary. Alternatively, if $n = a^2 + b^2$, then $a^2 + b^2 = 0 \ (mod\ 3)$, implying that $a^2 = 0 \ (mod\ 3)$ and $b^2 = 0 \ (mod\ 3)$, since for any integer x, we have $x^2 = 0$ or $1 \ (mod\ 3)$. Then $a = 0 \ (mod\ 3)$ and $b = 0 \ (mod\ 3)$, so $n = a^2 + b^2$ becomes $n = (3j)^2 + (3k)^2 = 9(j^2 + k^2) = 0 \ (mod\ 9)$, a contradiction.

SUMS OF FOUR OR FEWER SQUARES

By the corollary of the previous section, we see that there are infinitely many positive integers, such as 3^{2n+1}, which are not expressible as the sum of two squares. Some positive integers, such as $11 = 9 + 1 + 1$ and $21 = 16 + 4 + 1$, require three squares. Others, such as $7 = 4 + 1 + 1 + 1$ and $15 = 9 + 4 + 1 + 1$, require four squares. As we will see shortly, Lagrange proved (using several key ideas developed by Euler) that no positive integer requires more than four squares, that is, every positive integer can be written as the sum of four squares. (We are counting 0 as a square, or the last sentence would have to be changed to "every positive integer can be written as the sum of four or fewer squares.")

Example: Show that any positive integer of the form 3^{2n+1} is expressible as the sum of three squares. We have $3^{2n+1} = 3 \times 3^{2n} = (3^n)^2 + (3^n)^2 + (3^n)^2$.

Example: Show that if the positive integer k is expressible as the sum of three squares, then so is k^{2n+1}. Let $k = a^2 + b^2 + c^2$. Then $k^{2n+1} = k \times k^{2n} = (a^2 + b^2 + c^2)k^{2n} = (ak^n)^2 + (bk^n)^2 + (ck^n)^2$.

We will now see that there are infinitely many positive integers that are not expressible as the sum of three squares. This will require the following lemma.

Lemma: Let $n = 7$ (*mod* 8). Then n is not expressible as the sum of three squares.

Proof: For any integer, x, we have $x^2 = 0$, 1, or 4 (*mod* 8). Now, suppose that $n = a^2 + b^2 + c^2$. Then this becomes $7 = a^2 + b^2 + c^2$ (*mod* 8), which is clearly impossible. □

With the aid of this lemma, we have the following theorem.

Theorem: The positive integer $4^m(8k + 7)$ is not expressible as the sum of three squares.

Proof: Suppose, to the contrary, that $4^m(8k + 7) = a^2 + b^2 + c^2$, where $m \geq 1$. Then each of a, b, and c must be even. (To see this, note that $x^2 = 0$ or 1 (*mod* 4) and $4^m(8k + 7) = 0$ (*mod* 4). Thus, $a^2 + b^2 + c^2 = 0$ (*mod* 4) requires that $a^2 = b^2 = c^2 = 0$ (*mod* 4).) Then let $a = 2a'$, $b = 2b'$, and $c = 2c'$. Then $4^m(8k + 7) = a^2 + b^2 + c^2$ becomes $4^m(8k + 7) = (2a')^2 + (2b')^2 + (2c')^2$, which after simplification yields $4^{m-1}(8k + 7) = (a')^2 + (b')^2 + (c')^2$. Repeating this process $m - 1$ times yields a representation of $8k + 7$ as the sum of three squares, which contradicts the previous lemma. □

Example: Show that 240 is not expressible as the sum of three squares. Since $240 = 4^2 \times 15$, the claim is an immediate consequence of the above theorem.

The following lemma is due to Euler. It will be important in the proof of Lagrange's theorem that any positive integer is expressible as a sum of four squares.

Euler's Lemma: If two integers are expressible as the sum of four squares, so is their product.

Proof: Let $m = w^2 + x^2 + y^2 + z^2$ and let $n = s^2 + t^2 + u^2 + v^2$. Then $mn = a^2 + b^2 + c^2 + d^2$, where

$$a = ws + xt + yu + zv$$
$$b = wt - xs + yv - zu$$
$$c = wu - xv - ys + zt$$
$$d = wv + xu - yt - zs.$$
□

Corollary: Given finitely many integers, each of which is expressible as the sum of four squares, so is their product. □

The next lemma we shall require is interesting in its own right.

Lemma: Let p be an odd prime. Then the equation $x^2 + y^2 + 1 = 0$ (*mod p*) has a solution $x = a$ and $y = b$ such that $0 \le a \le \dfrac{p-1}{2}$ and $0 \le b \le \dfrac{p-1}{2}$.

Proof: Given the odd prime p,

let $A = \left\{ 1 + 0^2, 1 + 1^2, 1 + 2^2, \ldots, 1 + \left(\dfrac{p-1}{2}\right)^2 \right\}$

and let $B = \left\{ -0^2, -1^2, -2^2, \ldots, -\left(\dfrac{p-1}{2}\right)^2 \right\}$. Note that no two members of A are congruent. To see this, suppose that two members of A are congruent, that is, suppose $1 + a^2 = 1 + b^2$ (*mod p*). Then $a^2 = b^2$ (*mod p*), implying that either $a = b$ (*mod p*) or $a = -b$ (*mod p*). But the second possibility implies that $a + b = 0$ (*mod p*), which is impossible. A similar argument shows that no two members of B are congruent. Now, $A \cup B$ has $2[1 + (p-1)/2] = p + 1$ members, while there are only p incongruent numbers *mod p*. It follows that some member of A is congruent to some member of B, that is, $1 + a^2 = -b^2$ (*mod p*), or equivalently, $a^2 + b^2 + 1 = 0$ (*mod p*), where $0 \le a \le \dfrac{p-1}{2}$ and $0 \le b \le \dfrac{p-1}{2}$. $\square$

Example: Consider the odd prime 7. Then $A = \{1, 2, 5, 10\}$ and $B = \{0, -1, -4, -9\}$. These sets, reduced *mod 7*, become $A = \{1, 2, 5, 3\}$ and $B = \{0, 6, 3, 5\}$. Then $A \cap B = \{3, 5\}$. Using 3, we get $1 + 3^2 = -2^2$ (mod 7), or equivalently, $2^2 + 3^2 + 1 = 0$ (*mod 7*).

Corollary: Let p be an odd prime. Then there exists a positive integer $n < p$ such that np is the sum of four squares.

Proof: By the lemma, there exist integers a and b such that $0 \le a < \dfrac{p}{2}$ and $0 \le b < \dfrac{p}{2}$, and $a^2 + b^2 + 1 = np$ for some n. Now, $np = a^2 + b^2 + 1 < 2\left(\dfrac{p^2}{4}\right) + 1 < p^2$, implying that $n < p$. $\square$

The following lemma will be needed in the proof of the next theorem.

Lemma: Let n be an odd positive integer. Then given any integer x, there exists an integer y satisfying $|y| < \dfrac{n}{2}$ such that $x = y$ (*mod n*).

Proof: The integers $-\frac{n-1}{2}, -\frac{n-3}{2}, ..., -2, -1, 0, 1, 2, ..., \frac{n-3}{2}, \frac{n-1}{2}$ form a complete set of residues. Let y be a member of this set congruent to x. It follows that $|y| < \frac{n}{2}$.   □

Here is the final theorem that will enable us to obtain the result we have been anticipating.

Theorem: Any prime can be expressed as the sum of four squares.

Proof: Since the prime $2 = 1^2 + 1^2 + 0^2 + 0^2$, we need to consider only odd primes. Given the odd prime p, we invoke the previous corollary, and let n be the smallest positive integer such that np is the sum of four squares, that is, $np = a^2 + b^2 + c^2 + d^2$. By the preceding corollary, $n < p$. The proof will be over if we show that $n = 1$. The proof will be divided into two parts. In the first part, we will show that n is odd. In the second, we will show that $n = 1$.

Part 1: Assume that n is even. Then in the equation $np = a^2 + b^2 + c^2 + d^2$, the integers a, b, c, and d are either (i) all even, (ii) all odd, or (iii) two are even and two are odd. In all cases, we have, after possibly rearranging them, $a = b$ (*mod* 2) and $c = d$ (*mod* 2). We then have $\frac{np}{2} = \left(\frac{a-b}{2}\right)^2 + \left(\frac{a+b}{2}\right)^2 + \left(\frac{c-d}{2}\right)^2 + \left(\frac{c+d}{2}\right)^2$, where all summands on the right are integers. This contradicts the fact that n is the smallest integer such that np is the sum of four squares. Thus, n must be odd.

Part 2: We shall now prove that $n = 1$, by contradiction. Assume that $n \geq 3$. By the above lemma, there exist integers w, x, y, and z such that $|w| < \frac{n}{2}, |x| < \frac{n}{2}, |y| < \frac{n}{2},$ and $|z| < \frac{n}{2}$ and such that $w = a$ (*mod n*), $x = b$ (*mod n*), $y = c$ (*mod n*), and $z = d$ (*mod n*). Then $w^2 + x^2 + y^2 + z^2 = a^2 + b^2 + c^2 + d^2$ (*mod n*), implying that $w^2 + x^2 + y^2 + z^2 = mn$, for some nonnegative integer m. Now, $mn = w^2 + x^2 + y^2 + z^2 < 4\left(\frac{n}{2}\right)^2 = n^2$.

Thus, $m < n$.

Observe that m is positive. To see this, suppose to the contrary that $m = 0$. Then we would have $w^2 + x^2 + y^2 + z^2 = 0$, implying that $w = x = y = z = 0$, in which case it would follow that $a = b = c = d = 0$ (*mod n*). Then n^2 would divide each of a^2, b^2, c^2, and d^2, implying that $n^2 | a^2 + b^2 + c^2 + d^2$, or equivalently, that $n^2 | np$. This would imply that $n | p$, which is impossible since $1 < n < p$. So $1 \leq m < n$.

Now, $n^2mp = (mn)(np) = (w^2 + x^2 + y^2 + z^2)(a^2 + b^2 + c^2 + d^2)$. Then by Euler's Lemma, $n^2mp = A^2 + B^2 + C^2 + D^2$, where

$$A = wa + xb + yc + zd$$

$$B = wb - xa + yd - zc$$

$$C = wc - xd - ya + zb$$

$$D = wd + xc - yb - za.$$

Observe that n divides each of A, B, C, and D, since each of these is congruent, *mod n*, to $w^2 + x^2 + y^2 + z^2$. Then we deduce from $n^2mp = A^2 + B^2 + C^2 + D^2$ that $mp = \alpha^2 + \beta^2 + \gamma^2 + \delta^2$, where $\alpha = A/n$, $\beta = B/n$, $\gamma = C/n$, and $\delta = D/n$, which contradicts the fact that n is the smallest positive integer such that np is the sum of four squares.   □

We are finally ready to prove the theorem of Lagrange.

Theorem: Any positive integer can be expressed as the sum of four squares.

Proof: The result is trivial for $n = 1$. For $n > 1$, consider the prime decomposition of n into a product of primes (including repetition). Since, by the previous theorem, each prime is the sum of four squares, and since finite products of sums of four squares are sums of four squares, the result follows.   □

Given the positive integer n, consider $8n + 3$. Then $8n + 3 = a^2 + b^2 + c^2$. This becomes $3 = a^2 + b^2 + c^2$ (*mod 8*). Now, since $x^2 = 0$, 1, or 4 (*mod 8*), we must have $a = b = c = 1$ (*mod 8*). Then a, b, and c are odd, so $a = 2r + 1$, $b = 2s + 1$, and $c = 2t + 1$, in which case we obtain $8n + 3 = a^2 + b^2 + c^2 = (2r + 1)^2 + (2s + 1)^2 + (2t + 1)^2$, implying that $8n = 4r^2 + 4r + 4s^2 + 4s + 4t^2 + 4t$. This becomes, after dividing by 8 and factoring,

$$n = \frac{r(r+1)}{2} + \frac{s(s+1)}{2} + \frac{t(t+1)}{2}$$

proving the theorem of Gauss:

Theorem: Every positive integer can be expressed as the sum of three or fewer triangular numbers.   □

EXERCISES

An asterisk (∗) indicates that the exercise can be developed into a research project.

1 ∗Show that there are infinitely many solutions to the Diophantine equation $a^3 + b^3 = c^2$ given by $a = 6n^2 + 3n^3 - n$, $b = 6n^2 - 3n^3 + n$, and $c = 6n^2(3n^2 + 1)$. Then write a program to tabulate this for $n <= 50$.

2 ∗Ramanujan observed that 1729 is the smallest number that can be written as the sum of two cubes in two different ways, that is, $1729 = 729 + 1000 = 9^3 + 10^3$ and $1729 = 1 + 1728 = 1^3 + 12^3$. By computing the first 12 cubes, verify Ramanujan's observation. Then find five additional solutions.

3 ∗Find 50 solutions to the Diophantine equation $x^2 + y^2 = u^2 + v^2 = w^2$ such that $x \le y$, $u \le v$, and $x \ne u$. A solution to this Diophantine equation is equivalent to finding two Pythagorean triples with the same hypotenuse.

4 Write 13^5 as the sum of two squares.

5 Write 29^3 as the sum of two squares.

6 Show that 27 is the smallest number that can be written as the sum of three squares in two different ways. Find 100 more such numbers.

7 ∗Let A be the set of all (complex) numbers of the form $a + b\sqrt{-3}$, where a and b are integers. Find two different nontrivial ways to factor 4 in A.

8 Show that 3 times the sum of the first k odd numbers equals the sum of the next k odd numbers for $k \le 100$. When $k = 4$, for example, we have $3 \times (1 + 3 + 5 + 7) = 9 + 11 + 13 + 15$.

9 ∗A positive integer n has a *consecutive partition* of length k if there exist positive integers a and k, where $k \ge 2$, such that $n = a + (a+1) + (a+2) + \cdots + (a+k-1)$. Show that n has a consecutive partition if and only if either (i) n is triangular or (ii) n is the difference of two nonconsecutive triangular numbers. Write a program to do this.

10 ∗Let r, s, t, and u be distinct triangular numbers such that $r + s = t + u$. Show how to use this fact to produce a positive integer N with two distinct consecutive partitions. Then write a program to find such integers.

11 Show that the sum of k consecutive positive integers, where $k \geq 3$, is divisible by k if k is odd and is divisible by $\dfrac{k}{2}$ if k is even.

12 Show using the previous exercise that a prime number has no consecutive partition of length $k \geq 3$. Verify this for the first 20 primes.

13 Show that 2^n has no consecutive partition for $n \leq 10$.

14 Let $n = 2\,m$, where m is odd and $m > 3$. Show that n has a consecutive partition of length 4.

15 *A positive integer n has an *arithmetic partition* of length k if there exist positive integers a, k, and d such that $n = a + (a+d) + (a+2d) + \cdots + (a + (k-1)d)$. For obvious reasons, d is called the *difference*. When $d = 1$, we have a consecutive partition. If n has a consecutive partition of length k, show that any multiple of n has an arithmetic partition of length k. What's the difference? (That is, find d.)

16 Write a program to find all the arithmetic partitions of a given positive integer.

17 Find all solutions (if any) of the equation $x^2 + x = y^4 + y^3 + y^2 + y$ where x and y are less than 10.

18 Does the equation of the previous exercise have solutions if we remove the requirement that x and y must be positive?

19 List all the partitions of 20 and verify, by counting, that the number of distinct partitions equals the number of odd partitions. Remember $5 + 1$ is the same partition as $1 + 5$.

20 Show, by a correspondence chart, that the number of partitions of 20 consisting of 4 or fewer summands equals the number of partitions of 20 such that no summand exceeds 4.

21 *It is known that every positive integer has a *distinct* partition consisting of the Fibonacci numbers. Thus, for example, $10 = 2 + 3 + 5 = u_3 + u_4 + u_5$. Find such partitions for each of the first 20 positive integers.

22 Let $p(n)$ denote the number of partitions of n. Find $p(n)$ for $n = 1$, 2, ..., 20.

23 *Call a partition of n such that each summand is a power of 3 (including 1 that is 3^0) a *ternary partition*. If, in addition, each summand is distinct, call the partition a *distinct ternary partition*.

Show that every nonnegative integer has a ternary partition. Which nonnegative integers have *distinct* ternary partitions?

24 *Show that every nonnegative integer can be written as $a_0 + a_1 3 + a_2 3^2 + a_3 3^3 + a_4 3^4 + \cdots$, where each $a_n = 0$ or ± 1. For example, $2 = -1 + 3$, and $19 = 1 - 3^2 + 3^3$.

25 *Show that there are over 100 consecutive triples $a, a+1, a+2$, such that a is a square and each of $a+1$ and $a+2$ can be written as the sum of two squares. (The first triple with this property is 16, 17, 18.)

26 *A positive integer n has a *geometric partition* of length k if there exist positive integers a, k, and r, with $k \geq 3$, such that $n = a + ar + ar^2 + \cdots + ar^{k-1}$. For obvious reasons, r is called the *ratio*. If $a = 1$, the geometric partition is called *primitive*.

a. Show that $a \mid n$.

b. Show that if a prime number has a geometric partition, it must be primitive.

c. Show that $2^m - 1$, where $m \geq 3$, has a primitive geometric partition.

d. Let $n = 1 + r + r^2 + r^3 + \cdots + r^{k-2} + r^{k-1}$, where $k = 2j$ and $j \geq 3$. Show that n has a partition of length j and ratio r^2.

e. Let $n = 1 + r + r^2 + \cdots + r^{k-1}$, where $k = mj$ and $j \geq 3$. Show that n has a partition of length j and ratio r^m.

f. Let $n = 1 + r + r^2 + \cdots + r^{k-1}$, where $k = mj$ and $3 \leq m < j$. Show that $n = uv$, where u and v have primitive geometric partitions.

27 Write a program to find all the geometric partitions of a given positive integer. Use your program to find a geometric partition for 121 with initial term 1.

28 Show that the number of ordered partitions of n consisting of 1's, 2's, and 3's is given by v_n, where $v_1 = 1$, $v_2 = 2$, $v_3 = 4$, and $v_n = v_{n-1} + v_{n-2} + v_{n-3}$.

29 Write a program to find all solutions of the Diophantine equation $x^2 + y^2 = z^3$, where $1 \leq x \leq y < z \leq 1000$.

30 Find all solutions to the Diophantine equation $x^2 + y^2 = u^2 + v^2 = w^2$ such that $x \leq y$, $u \leq v$, $x \neq u$, and $w \leq 100$.

31 *Write p^3 as the sum of two squares for the first 10 primes p such that $p = 1 \ (mod \ 4)$.

32   Write a program to find all the ordered partitions of a given positive integer. (Here $5 + 1$ is different than $1 + 5$).

33   Modify the program of the previous exercise to find all the ordered partitions of a given positive integer such that:

   a. Each summand is odd.

   b. Each summand is distinct.

   c. Each summand does not exceed k, where k is a given positive integer.

   d. There are at most k summands, where k is a given positive integer.

34   Write a program to convert a given positive integer from decimal to binary form. Given 23, for example, the program should yield the binary number 10111, since $23 = 1 + 2 + 4 + 16$. *See programs at the end of the chapter.*

35   Write a program to find all the ways in which a given positive integer can be written as the sum of four or fewer squares.

36   Write a program to find all the ways in which a given positive integer can be written as the sum of three or fewer triangular numbers.

37   It is know that every integer is the sum of nine or fewer cubes. Write a program to find a partition for any given integer as the sum of the least number of cubes. Given 72, for example, your program should return the partition $64 + 8$, rather than a longer partition such as $27 + 27 + 8 + 8 + 1 + 1$ or $64 + 1 + 1 + 1 + 1 + 1 + 1 + 1 + 1$.

38   It is known that every fourth power is the sum of two triangular numbers. For example, $2^4 = 16 = 6 + 10 = t_3 + t_4$. Verify this for the next 10 fourth powers.

39   *The equation $x^2 + 4 = y^3$ has the solution $x = y = 2$. It is known that there exists only one more solution. Find it.

40   *Euler incorrectly conjectured that the Diophantine equation $u^5 = x^5 + y^5 + z^5 + w^5$ has no solution. Find a solution with $u = 144$.

41   Show that 325 is the smallest number that can be written as the sum of two squares in three different ways. Find the next number with this property.

Binary Representation Using Built-in to String Method

```
<!DOCTYPE HTML>
<html>
<meta charset="UTF-8">
<head>
<title>Decimal to Binary</title>
<script type="text/javascript"
src="primeDecomposition.js">
</script>
<script>

function convert(){
  var table;
  placeref = document.getElementById("place");
  n = Math.abs(parseInt(document.f.num.value));
  t = n.toString(2);
  messages = "The representation of "+ String(n)+"
in binary is "+ t;
  placeref.innerHTML = messages;
  return false;
}
</script>
</head>
<body>
Enter number to see its binary representation:
<form name="f" onSubmit="return convert();">
 <input type="number" name="num" value=""/>
 <input type="submit" value="Enter positive
integer"/>
</form>
<div id="place">
Answer will go here.
</div>
</body>
</html>
```

Binary Representation by Computation

```
<!DOCTYPE HTML>
<html>
```

```
<meta charset="UTF-8">
<head>
<title>Decimal to Binary Hard way</title>

<script>

function convert() {
  placeref = document.getElementById("place");
  placeref.innerHTML = "";
  n = Math.abs(parseInt(document.f.num.value));
  if (n==0) {
    message = "The representation of 0 in binary
is 0";
  }
  else {
    t = extractPowersOf2(n);
    message = "The representation of "+ String(n)+
" in binary is "+ t;
  }
  placeref.innerHTML = message;
  return false;
}
function extractPowersOf2(n) {
  var p;
  var rep;
  var q;
  //first phase: determining highest power
  p = 0;
  while(Math.pow(2,p)<n) {
    p++;
  }

  // p is the lowest power of 2 with 2 to the p >= n
  if (Math.pow(2,p)==n) {
    rep =  "1" + producenzeros(p);
  }
  else {
    p--;
    rep = "1" + producenzeros(p);
    q = 1; //needed to calculate position in the rep
// string
    n = n - Math.pow(2,p);
    p--;
```

```
    while (n>0) {
        if (n>=Math.pow(2,p)) {
           rep  = changeAtq(q,rep);
           n = n-Math.pow(2,p);
        }
        p--;
        q++;
    }


    }


   return rep;
}
function changeAtq(q,rep) {
      var nrep;
   nrep = rep.substring(0,q)+"1"+rep.substring
(q+1);
   //alert("nrep is "+nrep);
   return nrep;

}
function producenzeros(p) {
    var s = "";
    for (var i=0;i<p;i++) {
       s+="0";
    }
    return s;
}
</script>
</head>
<body>
Enter number to see its binary representation:
<form name="f" onSubmit="return convert();">
 <input type="number" name="num" value=""/>
 <input type="submit" value="Enter positive
integer"/>
</form>
<div id="place">
Answer will go here.
</div>
</body>
</html>
```

9

CRYPTOGRAPHY

We shall see how number theory plays a major role in encoding information, such as credit card numbers and banking transactions transmitted via the Internet.

INTRODUCTION AND HISTORY

In this chapter, we will examine several applications of number theory to *cryptography*, the science of encoding information. With the advent of electronic media such as the telephone and computer, it became necessary to encode military, diplomatic, corporate, and financial data. A simple credit card transaction, for example, transmitted from a store or bank to a financial institution must be encoded lest someone intercept the credit card number, expiration date, user name, and password. A more urgent example is afforded by the need to encrypt a message from the Pentagon to a Trident nuclear submarine containing instructions to relocate. The victory of the United States and Britain over Japan and Germany in World War II may be attributed in part to the breaking of the Japanese and German codes by American and British mathematicians and computer scientists such as Alan Turing. The United States

Elementary Number Theory with Programming, First Edition. Marty Lewinter and Jeanine Meyer.
© 2016 John Wiley & Sons, Inc. Published 2016 by John Wiley & Sons, Inc.

used the Navajo language in their radio transmissions in the Pacific Theater—a language unknown to the Japanese.

Here is a bit of standard terminology. The message to be *encrypted* (the technical term for *encoded*) is called *plaintext*, while the encrypted text is called the *ciphertext*. The ciphertext must be *decrypted* (the technical term for *decoded*) in order to be recovered as plaintext.

One of the earliest techniques was used by Julius Caesar in the first century B.C. It was an example of a *substitution code*, a code in which each letter is replaced by another. The encryption was accomplished, in terms of our alphabet today, by replacing each letter in the first row of the chart in Table 9.1 by the letter beneath it. The recipient decrypted the message by replacing each letter in the second row of the chart by the letter above it.

Notice that the substituted letter can be found without the chart by moving forward three letters. Thus, to find the replacement for G, we start at G and advance three letters: from G to H, from H to I, and from I to J. To encode the last three letters, X, Y and Z, we must "wrap around" the alphabet, that is, the successor of Z is A. Hence, Y, for example, is replaced by B, since we advance from Y to Z, from Z to A, and from A to B. Of course, there is nothing special about the decision to advance three letters.

By assigning the (two-digit) integers 01, 02, 03, ..., 24, 25, 26 to the alphabet, that is, A = 01, B = 02, ..., Z = 26, we can use modular arithmetic to accomplish the above encryption and decryption as follows. Letting p represent the assigned integer of a plaintext letter and letting c represent the integer of the resulting ciphertext letter, we have

$$c = p + 3 \,(mod\,26) \qquad\qquad (9.1)$$

The modular equation automatically does the "wrap around." Thus, for the plaintext letter X, we have $p = 24$, since X is the 24th letter of the alphabet. Then Equation 9.1 yields $c = 27 \,(mod\,26) = 1 = 01$, and the ciphertext letter is A, in agreement with our chart. To decipher the ciphertext, we use the modular equation

$$p = c - 3 \,(mod\,26) \qquad\qquad (9.2)$$

Note that we can replace (9.1) by any linear congruence equation $c = ap + b \,(mod\,26)$, provided that $gcd(a, 26) = 1$. In place of (9.2), we must then use $p = a'(c - b) \,(mod\,26)$, where a' is the multiplicative inverse of a $(mod\,26)$, that is, $aa' = 1 \,(mod\,26)$.

TABLE 9.1

A	B	C	D	E	F	G	H	I	J	K	L	M	N	O	P	Q	R	S	T	U	V	W	X	Y	Z
D	E	F	G	H	I	J	K	L	M	N	O	P	Q	R	S	T	U	V	W	X	Y	Z	A	B	C

TABLE 9.2

A	B	C	D	E	F	G	H	I	J	K	L	M	N	O	P	Q	R	S	T
01	02	03	04	05	06	07	08	09	10	11	12	13	14	15	16	17	18	19	20
U	V	W	X	Y	Z	;	.	?	0	1	2	3	4	5	6	7	8	9	!
21	22	23	24	25	26	27	28	29	30	31	32	33	34	35	36	37	38	39	40

Example: Encrypt the message *I LOVE MATH* using (9.1). We have I = 09, L = 12, O = 15, V = 22, E = 05, M = 13, A = 01, T = 20, and H = 08. Then the plaintext is 09 12152205 13012008. Using (9.1), we get the ciphertext 12 15182508 16042311.

Example: Decrypt the ciphertext 0904242222 using (9.2). Use (9.2) to decrypt the given ciphertext obtaining 0601211919, which becomes *GAUSS*, since G = 06, A = 01, U = 21, and S = 19.

One can make up one's own code by assigning a two-digit integer to each letter, punctuation symbol, and digit. A space between words is usually represented by "00." The example shown in Table 9.2 uses the two-digit numbers from 01 to 40.

Example: Encrypt the plaintext *EULER, GAUSS, AND FERMAT WERE GREAT!* using Table 9.2. We have

052112051827000701211919270001140400060518130120002305180500071805012040

Notice that we don't need to separate words since 00 denotes a space.

Example: Decrypt the ciphertext 0919003331000100161809130529. We obtain the plaintext *IS 31 A PRIME?*

Substitution codes are very easy to crack especially when they are used in lengthy messages. One reason is that letters occur with fairly consistent frequencies. The most frequent letter is E, for example, while Q and Z are rare. Substitution codes are, therefore, unsecure and no longer in use.

I came up with three programs for counting letters. (See the code at the end of the chapter.) All three use a similar interface. The html element for inputting the text is a `textarea` element together with a `form` element. The user can type in text or copy and paste. The whole entry may not be visible (though you can write code to expand the visible portion).

My code iterates through the whole text and makes use of an associative array to count occurrences. An associative array is an array that uses keys instead of numbers to index the values. The keys for these examples are the single character strings representing the letters. If a letter has not yet occurred in the text and so has not been entered into the array, I start it off with a value of 1; otherwise, my code increments the value. The code for this is

```
if (counts[letter]){
      counts[letter]++;
      }
    else {
      counts[letter]=1;
    }
```

To produce the results, I make use of a for-loop for associative arrays. The loop for my first example is

```
for(key in counts) {
  message += key + ": "+ String(counts[key])
+"<br/>";}
```

Now, one problem emerged: my code was counting blanks and punctuation! Since I didn't want this, for the first program, I created an array I named `notletters` and inserted code to see if something was not a letter, that is, on the list. For the second and third program, I took another approach. The code uses a variable I named `letters` that holds the alphabet. The `counts` array is initialized with entries with a key for each letter and a value of zero. The program has a simplified version of the loop for going through the text. The blanks and punctuation are still scanned, and something is done to entries in the `counts` array, though the statement `counts[txt[i]]++;` does nothing because nothing has been initialized to a number. I only care about the letters.

My code uses a regular for-loop to iterate through `letters` and extract the counts from the `counts` array to create an array of arrays, called `lettercount`, with one two-element array for each letter. For both the second and the third program, my code outputs table elements to display the results. For the third version of the program, I sort the array. JavaScript has a built-in `sort` method for arrays that allows the programmer to specify how sorting is to be done. The code

```
Lettercount = lettercount.sort(function(a,b){
      return a[1]<b[1];}   );
```

says *sort the elements of lettercount and do it by comparing the second (index is 1) element of each of the array elements.*

Recall that every integer can be written in binary form, that is, as a string of 1's and 0's representing the powers of 2 in the given integer's distinct binary partition. The *binary complement* of a binary number substitutes 0's for 1's and 1's for 0's. Thus, the binary complement of 11000101 is 00111010.

Now, plaintext (as in the example *IS 31 A PRIME?*) can be converted into a binary number M. Suppose the sender and receiver have a lengthy binary string N known only to them. Then M can be encrypted into a binary number M' as follows. If the ith digit (counted from the leftmost digit) of N is 1, let the ith digit of M' be the complement of the ith digit of M. If the ith digit of N is 0, let the ith digit of M' be the same as the ith digit of M. The receiver of the message uses the same scheme to decrypt M' into M, after which the agreed-upon table (say, Table 9.2) is used to recover the plaintext. This scheme works since the complement of the complement yields the original number. Note that we are assuming that N is longer than M. If this is not the case, we must break M into several smaller binary numbers.

Example: Use $N = 11001011110$ to encrypt $M = 10010101$. We show the work in the following chart. We get $M' = 01011110$.

N	1	1	0	0	1	0	1	1	1	1	0
M	1	0	0	1	0	1	0	1			
M'	0	1	0	1	1	1	1	0			

PUBLIC-KEY CRYPTOGRAPHY

A *public-key cryptosystem* is based on modular arithmetic. The idea is that people can send messages to other people by knowing each person's public key and using it to encrypt the message. The receiver decrypts the message using the private key. This involves a publicly known pair of integers, n and k, for each user chosen as follows. Let n be a large semi-prime, that is, let $n = pq$, where p and q are large primes chosen by the user. (We will soon say just how large these numbers must be.) The user then chooses a so-called encryption exponent, k, such that $gcd(k, \phi(n)) = 1$. Recall from Chapter 7 that $\phi(n) = \phi(pq) = \phi(p)\phi(q) = (p-1)(q-1)$. While n is public knowledge, p, q, and $\phi(n)$ are known only to the user.

Now let M be a plaintext formed using, say, Table 9.2, that is, M is a long string of two-digit numbers, such as 0919003331000100016 1809130529, representing a secret message. The sender then encrypts M by looking up the user's n and k and calculating $N = M^k \pmod{n}$.

The sender then transmits N to the user. The user figures out the so-called decryption exponent, j, from the modular equation $jk = 1 \pmod{\phi(n)}$, which can be solved since $gcd(k, \phi(n)) = 1$. We will use this fact in the equivalent formulation $jk = 1 + r\phi(n)$. Note that only the user can calculate j since this requires $\phi(n)$ that can't be found in "real time" without knowing the prime factors, p and q, of n. We are assuming that p and q are large enough to make the task of factoring n require years of running time on state-of-the-art supercomputers.

Knowing the decryption exponent, j, the user recovers M from N using the equation

$$N^j = \left(M^k\right)^j = M^{kj} = M^{1 + r\phi(n)} = M\left(M^{\phi(n)}\right)^r = M \times 1^r = M \pmod{n}$$

in which we used Euler's theorem, $M^{\phi(n)} = 1 \pmod{n}$. (See Chapter 7.) Of course, this requires $gcd(M, n) = 1$, which is virtually guaranteed to be the case unless either of the two prime factors of n (p and q) divide M. This cryptosystem was developed in 1977 by Rivest, Shamir, and Adleman (RSA). It is therefore known as the *RSA* system.

FACTORING LARGE NUMBERS

The RSA system rests on the difficulty of factoring large numbers. Fermat developed the following method of factoring large numbers.

If we wish to factor a large even number, n, we observe that 2 is a factor and therefore divide by 2 and consider $\frac{n}{2}$. If $\frac{n}{2}$ is even, divide it by 2, note that $4 = 2^2$ is a factor of n, and consider $\frac{n}{4}$. After k divisions by 2, we will reach an odd number, m, and realize that $n = 2^k m$. If $m = 1$, we are done. Otherwise, the problem of factoring n boils down to factoring the odd number m. So Fermat dealt exclusively with the problem of factoring a large *odd* number.

If $m = a^2 - b^2$, where $a - b > 1$, we have the nontrivial factorization $m = (a + b)(a - b)$. On the other hand, if m can be nontrivially factored as rs, then $m = \left(\frac{r+s}{2}\right)^2 - \left(\frac{r-s}{2}\right)^2$. Expand this equation and see that the right-hand side indeed reduces to rs. It is easy to see that $\frac{r+s}{2}$ and $\frac{r-s}{2}$ are integers, since r and s are both odd. (If r or s were even, m would be even.) Then their sum and difference must be even and, therefore, divisible by 2.

So Fermat developed a way to express m as the difference of two squares. That is, he tried to find a and b such that the given odd number $m = a^2 - b^2$ or $a^2 - m = b^2$. To do this, he found the smallest N for which $N^2 > m$. Then he considered the sequence $N^2 - m$, $(N+1)^2 - m$, $(N+2)^2 - m$, ..., until he found a square. This enabled him to write $m = a^2 - b^2 = (a+b)(a-b)$.

Example: Factor 119 using Fermat's method. The smallest N for which $N^2 > 119$ is 11. Now, $11^2 - 119 = 121 - 119 = 2$, which is not a square. We then examine $12^2 - 119 = 144 - 119 = 25$, which is a square. Then $119 = 144 - 25 = (12 - 5)(12 + 5) = 7 \times 17$.

Fermat's sequence $N^2 - m$, $(N+1)^2 - m$, $(N+2)^2 - m$, ..., must eventually yield a square, since $\left(\dfrac{m+1}{2}\right)^2 - m = \left(\dfrac{m-1}{2}\right)^2$. This ultimately yields the trivial factorization $m = 1 \times m$.

The interface for this program is set up like the others. The function `factor`, invoked by the form, first takes care of the trivial cases, 0, 1, and 2, and then handles even numbers. I did decide to make one function called `isEven` and another called `isaSquare`. The first is the single statement: `return (0==(n%2));` and the second is similar in that it returns a conditional expression:

```
function isaSquare(n) {
   var s = Math.sqrt(n);
   return (s==Math.floor(s));
}
```

You certainly can argue that the code in these very small functions could be part of the calling function, but it is good practice to use functions (or methods when building objects with methods and properties), and it makes the code more readable.

The workhorse of this program is the function `factorOdd`. It is called with a parameter that I named m. I decided to find the smallest number N whose square is bigger than the parameter and then proceed with testing the sequence $N^2 - m$, $(N+1)^2 - m$, etc. all in one for-loop. However, before getting to that, I needed to set up the bounds, the starting and stopping points for the loop. My thinking was that the loop would start at the `Math.floor(Math.sqrt(m))` and end with m itself.

When pondering that, it occurred to me that it made sense to check for the situation of m being a square separate from anything else, and so this is what I do. In my program (see the end of the chapter), I use variables a, b, as, bs, and rep. The a is the loop variable and corresponds to the N in the text. The as is calculated as the square of a. The bs is the difference: as − m. If bs is indeed a square, then b is calculated as the square root of bs. The variable rep is an array. The initial value is [1, m]. This is changed if a nontrivial factoring is found by the procedure.

Approximately 100 years after Fermat, Euler developed the following method of factoring odd numbers that can be written as the sum of two squares in two different ways. Let the odd number n satisfy

$$n = a^2 + b^2 = c^2 + d^2 \tag{9.3}$$

where a and c are odd and b and d are even. Without loss of generality, assume that $a > c$, in which case it follows that $d > b$. Then $a^2 - c^2 = d^2 - b^2$, implying that

$$(a-c)(a+c) = (d-b)(d+b) \tag{9.4}$$

Now, let $k = gcd[(a - c), (d - b)]$, so

$$a - c = kr$$
$$d - b = ks$$

where $gcd(r, s) = 1$. When these are inserted into (9.4), we obtain $r(a + c) = s(d + b)$. Since r and s are relatively prime, it follows that $s \mid a + c$. Then $a + c = st$. When this is substituted into $r(a + c) = s(d + b)$, we get $rst = s(d + b)$, yielding $d + b = rt$. So we have

$$a + c = st$$
$$d + b = rt$$

Observe that $t = gcd[(a + c), (d + b)]$. Since $a + c$ and $d + b$ are even, it follows that t is even. By similar reasoning, k is even.

Euler then obtains the following factorization of n:

$$n = \left[\left(\frac{k}{2}\right)^2 + \left(\frac{t}{2}\right)^2 \right] \left(r^2 + s^2\right) \tag{9.5}$$

To verify this, note that the right side of (9.5) becomes

$$\frac{1}{4}\left[(kr)^2 + (ks)^2 + (tr)^2 + (ts)^2\right] = \frac{1}{4}\left[(a-c)^2 + (d-b)^2 + (d+b)^2 + (a+c)^2\right]$$

$$= \frac{1}{4}\left(2a^2 + 2b^2 + 2c^2 + 2d^2\right) = \frac{1}{4}(2n + 2n) = n$$

Example: Let $n = 1{,}000{,}009 = 1000^2 + 3^2$. Euler found, using a table of squares, a second representation of n as the sum of two squares, $n = 972^2 + 235^2$. Then $a = 235$, $b = 972$, $c = 3$, and $d = 1000$. Then $a - c = 232$, $d - b = 28$, $a + c = 238$, and $d + b = 1972$. It follows that $k = gcd(232, 28) = 4$, from which we obtain $r = 58$ and $s = 7$. From $a + c = st$, we obtain $t = 34$. Then using (9.5), we have

$$1{,}000{,}009 = \left(2^2 + 17^2\right)\left(58^2 + 7^2\right) = 293 \times 3{,}413.$$

Much research has been devoted to the problem of factoring large numbers since the seventeenth century, and the current rapid rate of progress in the development of efficient factoring algorithms keeps the designers of cryptosystems on their toes.

THE KNAPSACK PROBLEM

Here is an interesting problem known as the *knapsack problem* that is used in cryptography. Given a set of distinct positive integers $S = \{n_1, n_2, \ldots, n_k\}$ and a number N, does N have a distinct partition consisting of members of S? If so, is the solution unique? Let's look at a few examples.

You can think of a knapsack as a linear container with capacity N and S the sizes of k items. A solution is a way of packing the knapsack with the items represented by S. Variations of the knapsack problem are categorized as NP-complete or NP-hard, indicating that so far, in spite of considerable efforts, no one has developed fast methods, that is, better methods than essentially trying everything. "Fast" in this context means an algorithm in which the time is bound by a polynomial in the size of the problem. This subject, computational complexity, is a great topic for readers of this book to pursue next!

Example: Let $S = \{1, 3, 5, 10, 15, 17\}$ and let $N = 18$. Then we have $18 = 17 + 1 = 15 + 3 = 10 + 5 + 3$, yielding three distinct partitions of 18 using members of S. On the other hand, for $N = 13$, we have the unique distinct partition $13 = 10 + 3$. Finally, when $N = 12$, there is no solution.

Example: Let $S = \{1, 2, 4, 8, 16, 32\}$. Then there is a unique distinct partition for each N satisfying $1 \le N \le 63$, namely, the unique binary partition implied by the base 2 expression for N.

The knapsack problem is presented as follows. Given the set of distinct positive integers $S = \{n_1, n_2, \ldots, n_k\}$ and the equation

$$N = a_1 n_1 + a_2 n_2 + a_3 n_3 + \cdots + a_k n_k \qquad (9.6)$$

is there a solution such that each a_i is either 1 or 0? To make this clear, observe that a solution $N = n_1 + n_4 + n_7$, for example, is equivalent to $a_1 = a_4 = a_7 = 1$, while all other a_i's are zero in (9.6).

When S is very large, it is difficult to determine whether a given N has a solution. This is because of the observation that if S has k members, then there are 2^k subsets of members of S. Furthermore, it is difficult to determine how many solutions there are for a given N after one solution has been found. Fortunately, if S is a special kind of sequence, this is not the case, as we shall now see.

SUPERINCREASING SEQUENCES

A sequence $S = \{n_1, n_2, \ldots, n_k\}$ is called *superincreasing* if

$$n_2 > n_1$$
$$n_3 > n_1 + n_2$$
$$n_4 > n_1 + n_2 + n_3$$
$$\vdots$$
$$n_k > n_1 + n_2 + n_3 + n_4 + \cdots + n_{k-1}$$

or, equivalently, if each term strictly exceeds the sum of all of its predecessors. The following algorithm solves the knapsack problem when S is a superincreasing sequence. We will see that for any given N, either (i) there is a unique solution or (ii) there is no solution.

1. Given N, find the largest n_i in S such that $n_i \le N$. If no such n_i exists, the algorithm terminates and there is no solution. If $n_i = N$, the algorithm terminates and we have a unique solution. If $n_i < N$, proceed to the next step. Observe that n_i must be included in a solution since the sum of all the predecessors of n_i is strictly less than n_i and, hence, strictly less than N.

2. Find the greatest n_j, where $j < i$, such that $n_j \le N - n_i$. This implies that $n_j + n_i \le N$. If no such n_j exists, the algorithm terminates and there is no solution. If $n_j + n_i = N$, the algorithm terminates and we have a unique solution. If $n_j + n_i < N$, proceed to the next step. n_j must be included in a solution since the sum of all the predecessors of n_j is strictly less than n_j.

3. Continue in this manner until either a solution is found, in which case it is unique, or the algorithm can't be continued, in which case there is no solution.

Example: Solve (9.6) for $N = 14$ where S is the superincreasing sequence $\{1, 2, 4, 8, 16\}$. Since 8 is the largest member of S not exceeding 14, it must be included. Now, $14 - 8 = 6$ and 4 is the largest member of S not exceeding 6, so it must also be included. 2 is the largest member of S not exceeding $14 - 8 - 4 = 2$, so 2 must also be included. Since $14 = 8 + 4 + 2$, the algorithm terminates and the solution is unique. 14 in base 2 is 1110 since $14 = 8 + 4 + 2 = 2^3 + 2^2 + 2^1$.

Example: Solve (9.6) for $N = 10$ where S is the superincreasing sequence $\{3, 4, 8, 16\}$. Since 8 is the largest member of S not exceeding 10, it must be included. Now, $10 - 8 = 2$, so the algorithm can't be continued and there is no solution.

Here is a cryptosystem based on the knapsack problem. A user selects a superincreasing sequence $n_1, n_2, \ldots, n_r$ and a modulus m such that m exceeds the sum of the n_i's. (For convenience, r should be a very large number. We will see why shortly.) Then the user selects a positive number a satisfying $gcd(a, m) = 1$ and $a < m$. Let a' satisfy $aa' = 1 \ (mod \ m)$, that is, a' is the multiplicative inverse of $a \ (mod \ m)$. Such an a' exists since $gcd(a, m) = 1$.

Now, the user determines the sequence of positive numbers $k_1, k_2, \ldots, k_r$ by solving the r equations $k_i = an_i \ (mod \ m)$, where $i = 1, 2, \ldots, r$, such that each $k_i < m$. Finally, the user publishes the sequence $k_1, k_2, \ldots, k_r$. This new sequence does not have the superincreasing property. (In the incredibly rare event that it does, repeat the above procedure using a different a.)

To send a message to our user, convert the text into a long binary string M of length, say, t, using a binary version of Table 9.2. That is, convert each number in Table 9.2 to its binary equivalent using a fixed number of digits for all entries. The letter C, for example, becomes 0000011 if we use seven digits for each binary number, while an empty space (formerly 00) becomes 0000000. Denoting the t digits of M by a_1, $a_2, a_3, \ldots, a_t$, calculate the number

$$N = a_1 k_1 + a_2 k_2 + a_3 k_3 + \cdots + a_t k_t \tag{9.7}$$

where, hopefully $t \le r$, the length of the sequence $k_1, k_2, \ldots, k_r$. If $r > t$, break M into smaller numbers. Note that the recovery of M from N involves solving the knapsack problem represented by Equation 9.7.

The sender transmits N to the user. If an unauthorized person intercepts the transmission, he would not be able to recover M from N since the knapsack problem (9.7), involves a sequence that is not superincreasing. The user, knowing m and a', converts (9.7) into a superincreasing knapsack problem by first multiplying both sides by a' and reducing $mod\ m$ as follows:

$$N' = a'N = a_1 a' k_1 + a_2 a' k_2 + a_3 a' k_3 + \cdots + a_t a' k_t (mod\ m)$$

where $N' < m$. This then becomes, using $k_i = a n_i\ (mod\ m)$ and $aa' = 1\ (mod\ m)$,

$$N' = a_1 a' a n_1 + a_2 a' a n_2 + a_3 a' a n_3 + \cdots + a_t a' a n_t = a_1 n_1 + a_2 n_2 +$$
$$a_3 n_3 + \cdots + a_t n_t (mod\ m)$$

This yields the knapsack problem $N' = a_1 n_1 + a_2 n_2 + a_3 n_3 + \cdots + a_t n_t$, which is easy to solve since the sequence is superincreasing. The user readily recovers the string of binary digits $a_1 a_2 a_3 \ldots$ comprising M.

The knapsack cryptosystem was developed in 1978 by Merkle and Hellman. It required modification several years later after Shamir developed an algorithm that solved the "encrypted" knapsack problem, that is, one using a sequence obtained from a superincreasing sequence $n_1, n_2, \ldots, n_r$ using the encryption equation $k_i = a n_i\ (mod\ m)$ with $gcd(a, m) = 1$.

EXERCISES

An asterisk ($*$) indicates that the exercise can be developed into a research project.

1 If a substitution code uses the equation $c = p + 13$ (*mod* 26) to encrypt letters of the alphabet, explain why this equation may be used to decipher these letters if we interchange c and p.

2 Use the encryption technique of the previous exercise to encrypt the plaintext NUMBER THEORY IS A GREAT SUBJECT.

3 *Show that the linear modular equation $c = 3p$ (*mod* 26) yields a valid substitution code, while $c = 2p$ (*mod* 26) does not.

4 Use the encryption technique of the previous exercise to encrypt the plaintext CRYPTOGRAPHY IS INTERESTING.

5 Show that the modular equation $c = p^2$ (*mod* 26) does not yield a valid substitution code.

6 *Generalize the previous exercise by showing that $c = p^k$ (*mod* 26), where k is even, does not yield a valid substitution code.

7 *Show that the cubic modular equation $c = p^3$ (*mod* 26) does not yield a valid substitution code.

8 Create a small-scale RSA code and use it to encrypt and decrypt the plaintext MATHEMATICS ENRICHES OUR LIVES.

9 *Repeat the previous exercise using a code based on the knapsack problem.

10 Let $n_1, n_2, \ldots, n_r$ be a sequence of positive integers for which $n_i > 2n_{i-1}$, for each $i = 2, 3, \ldots, r$. Show that this sequence is superincreasing.

11 Use Fermat's factoring method to factor 135.

12 Use Fermat's factoring method to factor 192 without reducing the problem to factoring the odd number obtained by repeatedly factoring out 2.

13 Why must an odd number be of the form $4k + 1$ in order for it to be factorable using Euler's method?

14 Explain why it follows from the previous exercise that a prime of the form $4k + 1$ cannot be written as the sum of two squares in two different ways.

15 Write a program that encrypts plaintext using the substitution code given by the linear congruence equation $c = 3p + 5$ (*mod* 26).

16 Write a program to decrypt ciphertext encrypted using the method of the previous exercise.

17 Write a program that converts plaintext into a number using Table 9.1.

18 Write a program that converts a number obtained using Table 9.1 into its original plaintext.

19 Write a program that, when given a lengthy text, prints out a table listing the frequencies for each letter of the alphabet used in the given text. Then input a text containing several thousand words. *See several programs at the end of the chapter.*

20 Write a program that factors odd numbers using Fermat's method. *See program at the end of the chapter. The program works for any number, that is, it returns a factoring for even numbers and returns the factoring 1 × m for numbers for which the procedure fails.*

21 Modify the program of the previous exercise to enable the user to input an even number. The program should begin by attempting to factor out 2 until an odd number is obtained.

Count Letters

```
<!DOCTYPE HTML>
<html>
<meta charset="UTF-8">
<head>
<title>Count Letters</title>
<script>
var txt;
var counts= new Array();
var notLetters = " .,;?/!()'-";
function countLetters(){
  placeref = document.getElementById("place");
  placeref.innerHTML = "";
  txt = document.f.mytext.value;
  for (var i=0;i<txt.length;i++){
    letter = txt[i];
    if (okletter(letter)){
      if (counts[letter]){
        counts[letter]++;
        }
      else {
        counts[letter]=1;
```

```
      }
    }
  }
  message = "";
  for(key in counts) {
      message += key + ": "+ String(counts[key])
+"<br/>";
  }
  placeref.innerHTML = message;
  return false;
}

function okletter(p) {
  if (notLetters.indexOf(p)>=0){
    return false;
  }
  else {
    return true;
  }
}
</script>
</head>
<body>
Counting letters: <br/>
 <textarea name="mytext" form="f">Enter text here
</textarea>
 <br/>
<form name="f" id="f" onSubmit="return
countLetters();">

 <input type="submit" value="Submit"/>
</form>
<div id="place">
Answer will go here.
</div>
</body>
</html>
```

Count Letters and Produce Table in Alphabetical Order

```
<!DOCTYPE HTML>
<html>
```

```html
<meta charset="UTF-8">
<head>
<title>Count Letters</title>

<script>
var txt;
var counts= new Array();
var letters = "abcdefghijklmnopqrstuvwxyz";
function countLetters(){
  placeref = document.getElementById("place");
  placeref.innerHTML = "";
  txt = document.f.mytext.value;
  txt = txt.toLowerCase();
  //initialize counts for the letters
  for (var i=0;i<letters.length;i++){
    counts[letters[i]] = 0;
  }
  for (var i=0;i<txt.length;i++){

    counts[txt[i]]++;
  }
  message = "<br/> <table border='1'>";
  for(var i=0;i<letters.length;i++){

message+="<tr><td>"+letters[i]
+"</td><td>"+String(counts[letters[i]])
+"</td></tr>";
  }
  message+="</table>";
  placeref.innerHTML = message;
  return false;
}

</script>
</head>
<body>
Counting letters: <br/>
 <textarea name="mytext" form="f">Enter text here
</textarea>
 <br/>
<form name="f" id="f" onSubmit="return
countLetters();">
```

```
<input type="submit" value="Submit"/>
</form>
<div id="place">
Answer will go here.
</div>
</body>
</html>
```

Count Letters and Produce Table in Order of Frequency

```
<!DOCTYPE HTML>
<html>
<meta charset="UTF-8">
<head>
<title>Count Letters & Sort</title>
<script>
var txt;
var counts= new Array();
var letters = "abcdefghijklmnopqrstuvwxyz";
var lettercount = [];
function countLetters(){
  placeref = document.getElementById("place");
  placeref.innerHTML = "";
  txt = document.f.mytext.value;
  txt = txt.toLowerCase();
  //initialize counts for the letters
  for (var i=0;i<letters.length;i++){
    counts[letters[i]] = 0;
  }
  for (var i=0;i<txt.length;i++){

    counts[txt[i]]++;
  }

// now create new array with letters and count
  lettercount = [];
  for(var i=0;i<letters.length;i++){
    lettercount.push([letters[i],counts
[letters[i]]]);
  }
  //now sort based on the second (1st index) value
```

```
   lettercount = lettercount.sort(function(a,b){
      return a[1]<b[1];
   });

   message="<br/> <table border='1'>";
   for (var i=0;i<lettercount.length;i++){

message+="<tr><td>"+lettercount[i][0]
+"</td><td>"+lettercount[i][1]+"</td></tr>";
   }
   message += "</table>";
   placeref.innerHTML = message;
   return false;
}

</script>
</head>
<body>
Counting letters: <br/>
 <textarea name="mytext" form="f">Enter text here
</textarea>
 <br/>
<form name="f" id="f" onSubmit="return
countLetters();">

 <input type="submit" value="Submit"/>
</form>
<div id="place">
Answer will go here.
</div>
</body>
</html>
```

Fermat Factoring

```
<!DOCTYPE HTML>
<html>
<meta charset="UTF-8">
<head>
<title>Fermat Factoring</title>
<script>
```

```
function factor() {
  placeref = document.getElementById("place");
  placeref.innerHTML = "";
  n = Math.abs(parseInt(document.f.num.value));
  if (n<=2) {
    message = "0, 1, 2 have no non-trivial factors";
  }
  else if (isEven(n)) {
    message = "A factoring of "+String(n)+" is "+
String(2)+" X "+ String(n/2);
  }
  else {
    t = factorOdd(n);
    message = "A factoring of "+ String(n)+" is "+ t
[0] + " X "+ t[1];
  }
  placeref.innerHTML = message;
  return false;
}
function factorOdd(m) {
    var a;
    var b;
    var a0 = Math.floor(Math.sqrt(m));
    var as;
    var b;
    var bs;
    var rep = [1,m];
    // need to take care of obvious case of m being a
// square itself
    if ((a0*a0)==m) {
      rep = [a0,a0];
    }
    else {
    for (a=a0+1;a<=m;a++) {
      //check if a*a bigger than m and then …
      as = a*a;
      if (as>m) {
        bs = as-m;
        if (isaSquare(bs)) {
          b = Math.sqrt(bs);
          rep = [a+b,a-b];
          break;
```

```
        }
      }
    }
  }
    return rep;
}

function isEven(n){
 return (0==(n%2));

}

function isaSquare(n){
  var s = Math.sqrt(n);
  return (s==Math.floor(s));
}

</script>
</head>
<body>
Enter number to seek factoring:
<form name="f" onSubmit="return factor();">
 <input type="number" name="num" value=""/>
 <input type="submit" value="Enter positive
integer"/>
</form>
<div id="place">
Answer will go here.
</div>
</body>
</html>
```

ANSWERS OR HINTS TO SELECTED EXERCISES

CHAPTER 1

3. $\dfrac{1}{t_k} = \dfrac{2}{k(k+1)} = 2\left(\dfrac{1}{k} - \dfrac{1}{k+1}\right)$, so $\displaystyle\sum_{k=1}^{n}\dfrac{1}{t_k} = 2\sum_{k=1}^{n}\left(\dfrac{1}{k} - \dfrac{1}{k+1}\right) =$
$2\left(1 - \dfrac{1}{n+1}\right)$

4a. $\dfrac{a}{b} + \dfrac{c}{d} = \dfrac{ad+bc}{bd}$

9. Exactly one of n and $n+1$ is even. Divide that factor by 2 in the denominator of $\dfrac{n(n+1)}{2}$.

17. $1 + 2 + 2^2 + 2^3 + \cdots + 2^{n-1} = 2^n - 1 < 2^n$.

Elementary Number Theory with Programming, First Edition. Marty Lewinter and Jeanine Meyer.
© 2016 John Wiley & Sons, Inc. Published 2016 by John Wiley & Sons, Inc.

CHAPTER 2

2. $6k$, $6k+2$, $6k+3$, $6k+4$ are composite.

4. $\displaystyle\sum_{k=1}^{n}(2k-1)^2 = \sum_{k=1}^{n}4k^2 - 4k + 1 = 4\sum_{k=1}^{n}k^2 - 4\sum_{k=1}^{n}k + \sum_{k=1}^{n}1.$

6. Assume that the sum, $f(n)$, of the first n fourth powers is $an^5 + bn^4 + cn^3 + dn^2 + en + f$. Then find six equations by plugging in the values 1, 2, 3, 4, 5, and 6 for n.

13. 21 and 55.

14. Use the fact that the sum of two odd numbers is even and the sum of an odd number and an even number is odd. Then realize that the first two Fibonacci numbers are odd.

CHAPTER 3

5. For any positive integer m, we have $m! = m(m-1)(m-2)...(3)(2)(1) = m(m-1)!$. Now, let $m = n!$ in which case $m-1 = n! - 1$. Then $(n!)! = n!(n!-1)!$.

11. $(n)1 + (n-1)2 + (n-2)3 + (n-3)4 + \cdots + 3(n-2) + 2(n-1) + 1(n) = 1n + (2n-2) + (3n-6) + (4n-12) + \cdots + [nn - \{n(n-1)\}]n = (1 + 2 + 3 + 4 + \cdots + n)n +$ (the sum of the first $n-1$ oblong numbers). Then use the fact that the sum of the first n oblong numbers is $\dfrac{n(n+1)(n+2)}{3}$. Finally, $\dbinom{n+2}{3} = \dfrac{n(n+1)(n+2)}{6}$.

CHAPTER 4

1. $gcd(80,540) = gcd(60,80) = 20$.

2. $n^3 - n = (n-1)n(n+1)$. One of any three consecutive numbers is divisible by 3, and one of any two consecutive numbers is divisible by 2.

4. If m and n are odd, $m^2 + n^2 = (2a+1)^2 + (2b+1)^2 = 4a^2 + 4a + 1 + 4b^2 + 4b + 1 = 4k + 2$, where $k = a^2 + a + b^2 + b$.

8. $3^n + 1 = (4-1)^n + 1$. Now, use the binomial theorem to evaluate $(4-1)^n$ and note that when n is odd, it ends in -1, which is canceled by the 1 in $(4-1)^n + 1$. Then realize that each remaining term is divisible by 4.

14. First write x as $n + y$, where n is the integer $\lfloor x \rfloor$ and y satisfies $0 \le y < 1$. (For example, $6.7 = 6 + .7$) Then observe that $10x = 10n + 10y$. Now, show that $\lfloor 10x \rfloor = 10n + \lfloor 10y \rfloor$. Finally, figure out how big $\lfloor 10y \rfloor$ can get.

18. Find the exponents of the primes 2, 3, 5, 7, 11, 13, 17, 19, 23, and 29. The exponent 2, for example, is $\left\lfloor \dfrac{30}{2} \right\rfloor + \left\lfloor \dfrac{30}{4} \right\rfloor + \left\lfloor \dfrac{30}{8} \right\rfloor + \left\lfloor \dfrac{30}{16} \right\rfloor = 15 + 7 + 3 + 1 = 26$.

23. The difference of two odd numbers is even.

CHAPTER 5

1. $(3k-1)^3 = 27k^3 - 27k^2 + 9k - 1 = -1 \ (mod\ 9)$, $(3k)^3 = 27k^3 = 0 \ (mod\ 9)$, and $(3k+1)^3 = 27k^3 + 27k^2 + 9k + 1 = 1 \ (mod\ 9)$. This proves that $n^3 = 0$ or $\pm 1 \ (mod\ 9)$ and that the observed pattern persists.

3. It suffices to verify the statement for $n = 0, 1, 2, 3,$ and 4. Alternatively, we can use Fermat's little theorem with $p = 5$.

11. $3^2 = 4^2 \ (mod\ 7)$, while $3 \ne 4 \ (mod\ 7)$.

12. $x = 1, 3, 5,$ and 9.

18. $2^{100} = (2^3)^{33} \times 2 = 2 \ (mod\ 7)$.

CHAPTER 6

2. $\tau(n) = 6 = 2 \times 3$ implies that either $n = pq^2$ or $n = r^5$, where $p, q,$ and r are primes such that $p \ne q$. Now, $p = 3$ and $q = 2$ yields $n = 12$, $p = 2$ and $q = 3$ yields $n = 18$, while $r = 2$ yields $n = 32$. The minimum n is 12.

7. By definition, $\sigma(p^3) = p^3 + p^2 + p + 1$. Using formula (6.2), we have
$$\sigma(p^3) = \frac{p^4 - 1}{p - 1} = \frac{(p^2 - 1)(p^2 + 1)}{p - 1} = (p+1)(p^2 + 1) = p^3 + p^2 + p + 1.$$

9. $f(n) = f(1n) = f(1)f(n)$, so $f(1) = 1$.

14. Let $F(n) = \displaystyle\sum_{d|n} \mu(d)f(d)$. Then F is multiplicative. Now, $F(p^k) = \mu(1)f(1) + \mu(p)f(p) = f(1) - f(p) = 1 - f(p)$.

CHAPTER 7

1. $\phi(10^n) = 10^n \left(1 - \dfrac{1}{2}\right)\left(1 - \dfrac{1}{5}\right) = 4 \times 10^{n-1}$.

   $\phi(10!) = 10! \times \left(1 - \dfrac{1}{2}\right)\left(1 - \dfrac{1}{3}\right)\left(1 - \dfrac{1}{5}\right)\left(1 - \dfrac{1}{7}\right) = \left(\dfrac{8}{5 \times 7}\right) 10! = 2 \times$
   $3 \times 4 \times 6 \times 8^2 \times 9 \times 10 = 829440$.

2a. $\phi(3^n) = 3^n \left(1 - \dfrac{1}{3}\right) = 3^n \times \dfrac{2}{3}$.

8. $\phi(n^k) = n^k \left(1 - \dfrac{1}{p_1}\right)\left(1 - \dfrac{1}{p_2}\right) \cdots \left(1 - \dfrac{1}{p_r}\right)$.

9. $3^4 = 1$ (*mod* 10).

CHAPTER 8

7. $4 = 2 \times 2$ and $4 = \left(1 + \sqrt{-3}\right)\left(1 - \sqrt{-3}\right)$.

8. $(2k)^2 - k^2 = 3k^2$.

9. If $a = 1$, the sum is triangular. If not, the sum is $t_{a+k-1} - t_{a-1}$.

14. $n = \dfrac{m-3}{2} + \dfrac{m-1}{2} + \dfrac{m+1}{2} + \dfrac{m+3}{2}$.

15. Let $n = a + (a+1) + (a+2) + \cdots + (a + (k-1))$. Then multiplying both sides by m yields $nm = am + (am + m) + (am + 2m) + \cdots + (am + (k-1)m)$. Note that $d = m$, while the length of the partition is unchanged.

CHAPTER 9

1. $p = c - 13 = c + 13$ (*mod* 26).

3. Given $c = 3p$ (*mod* 26), we have $9c = 27p = p$ (*mod* 26), that is, $p = 9c$ (*mod* 26).

5. $(26 - p)^2 = p^2$ (*mod* 26).

7. Consider $p = 1$ and $p = 3$.

10. Use induction to show that $n_k > n_1 + n_2 + n_3 + \cdots + n_{k-1}$, for each $k = 2, 3, \ldots, r$. The base case is obvious since $n_2 > 2n_1 > n_1$. Now, assume that $n_k > n_1 + n_2 + n_3 + \cdots + n_{k-1}$. Then $n_{k+1} > 2n_k = n_k + n_k > n_1 + n_2 + n_3 + \cdots + n_{k-1} + n_k$.

INDEX

Note: Page numbers in *italics* refer to Figures; those in **bold** to Tables.

Elementary Number Theory with Programming, First Edition. Marty Lewinter and Jeanine Meyer.
© 2016 John Wiley & Sons, Inc. Published 2016 by John Wiley & Sons, Inc.

Printed and bound by CPI Group (UK) Ltd, Croydon, CR0 4YY

27/10/2024

14580269-0005